Freeze Drying of Pharmaceutical Products

Advances in Drying Science and Technology

Series Editor
Arun S. Mujumdar
McGill University, Quebec, Canada

For more information about this series, please visit: www.crcpress.com/Advances-in-Drying-Science-and-Technology/book-series/CRCADVSCITEC

Freeze Drying of Pharmaceutical Products

Edited by
Davide Fissore, Roberto Pisano, and
Antonello Barresi

CRC Press
Taylor & Francis Group
Boca Raton London New York

CRC Press is an imprint of the
Taylor & Francis Group, an **informa** business

CRC Press
Taylor & Francis Group
6000 Broken Sound Parkway NW, Suite 300
Boca Raton, FL 33487–2742

First issued in paperback 2021

ISBN-13: 978-0-367-07680-1 (hbk)
ISBN-13: 978-1-03-208590-6 (pbk)

Library of Congress Cataloging-in-Publication Data

Names: Fissore, Davide, editor. | Pisano, Roberto (Of Politecnico di Torino), editor. | Barresi, Antonello, editor.
Title: Freeze drying of pharmaceutical products / edited by Davide Fissore, Roberto Pisano, Antonello Barresi.
Other titles: Advances in drying science & technology.
Description: Boca Raton : CRC Press, [2020] | Series: Advances in drying science and technology | Includes bibliographical references and index. | Summary: "Pharmaceuticals typically rely on compounds that are unstable as aqueous solutions and therefore need to be dried for an extended shelf lifetime. While lyophilization remains the gold standard in dehydration technology, it is among the most expensive and time-consuming unit operations in pharma manufacturing. This book provides an overview of the most recent and cutting edge developments and technologies in the field, focusing on formulation developments and process monitoring and considering new technologies for process development like the micro freeze dryer. Case studies from freeze dryer manufacturer and pharmaceutical companies are discussed"—Provided by publisher.
Identifiers: LCCN 2019026901 (print) | ISBN 9780367076801 (hardback ; alk. paper) | ISBN 9780429022074 (ebook)
Subjects: MESH: Technology, Pharmaceutical | Freeze Drying—methods | Freeze Drying—trends | Pharmaceutical Preparations
Classification: LCC RM301.25 (print) | LCC RM301.25 (ebook) | NLM QV 778 | DDC 615.1/9—dc23
LC record available at https://lccn.loc.gov/2019026901
LC ebook record available at https://lccn.loc.gov/2019026902

Visit the Taylor & Francis Web site at
www.taylorandfrancis.com

and the CRC Press Web site at
www.crcpress.com

Contents

Series Preface

It is well known that the unit operation of drying is a highly energy-intensive operation encountered in diverse industrial sectors ranging from agricultural processing, ceramics, chemicals, minerals processing, pulp and paper, pharmaceuticals, coal polymers, food, forest products industries as well as waste management. Drying also determines the quality of the final dried products. Making drying technologies sustainable and cost effective via application of modern scientific techniques is the goal of academic as well as industrial R&D activities around the world.

Drying is a truly multi- and interdisciplinary area. Over the last four decades the scientific and technical literature on drying has seen exponential growth. The continuously increasing interest in this field is also evident from the success of numerous international conferences devoted to drying science and technology.

The establishment of this new series of books entitled *Advances in Drying Science and Technology* is designed to provide authoritative and critical reviews and monographs focusing on current developments as well as future needs. It is expected that books in this series will be valuable to academic researchers as well as industry personnel involved in any aspect of drying and dewatering.

The series will also encompass themes and topics closely associated with drying operations, for example, mechanical dewatering, energy savings in drying, environmental aspects, life cycle analysis, technoeconomics of drying, electrotechnologies, control and safety aspects, and so on.

ABOUT THE SERIES EDITOR

Arun S. Mujumdar is an internationally acclaimed expert in drying science and technologies. In 1978 he became the founding chair of the International Drying Symposium (IDS) series and has been editor-in-chief of *Drying Technology: An International Journal* since 1988. The fourth enhanced edition of his *Handbook of Industrial Drying*, published by CRC Press, has just appeared. He is recipient of numerous international awards including honorary doctorates from Lodz Technical University, Poland, and University of Lyon, France. Please visit www.arunmujumdar.com for further details.

Preface

FROM THE GLACIAL ERA TO IOT: THE LYOPHILIZATION GRAND ADVENTURE

Lyophilization is naturally occurring in nature, even if it is quite a rare phenomenon, as it requires extreme conditions: low temperature, an energy supply and an extremely dry environment (very low water partial pressure). Several anecdotal examples can be reported: from lithopanspermia theory, with living cells quite certainly freeze-dried while travelling space on rock fragments until they hit Earth, to Greenland mummies. In the preface to his book's second edition, Rey mentions mammoth meat from northern Siberia (offered in 1902 at the banquet of the International Congress of Paleontology) and reports of a big "mammoth steak with its fur still attached to the skin," received in the late 1960s from the USSR Academy of Science, which turned out not to be frozen, but naturally and perfectly freeze-dried: the animal was buried under snow, relatively close to the surface, where it was sublimed for millennia. In recent times, it has been reported that the poor quality boots of the Italian Alpine Corp in Russia during World War II underwent freeze-drying, losing mechanical properties, because of the extremely dry environment.

In previous centuries freeze-drying was employed by some populations for food conservation: the origins of this process are generally traced back to the 15th century in Peru, where Incas were freezing tubers and potatoes above Macchu Picchu, leaving them to be sublimated by the effects of the sun's radiation in the dry and low-pressure environment. Japanese monks on Mount Koya, south of Osaka, similarly prepared "kodayofu", preserved tofu, by packing bean curd into the snowy mountainside during the Kamakura period (AD 1185 to 1333), and Vikings in northern Europe used triangular wooden racks to freeze-dry codfish.

Despite these early applications, this technique for food preservation was forgotten. It appeared again as an innovation in food technology in 1955, and it is currently a technique used by some renowned chefs in exclusive restaurants or to prepare food for astronauts.

Much more recent is the use of lyophilization in the pharmaceutical and medical fields. Here, we summarize its use in these fields from early applications to current market perspectives.

Lyophilization was substantially unknown in Western countries until the early 1900s: the first attempt to lyophilize a biological product was documented by the German histologist Richard Altmann in 1881 (Altmann 1894), but only in the 1920s did lyophilization become an established laboratory tool for preserving live microorganisms or tissues (Jennings 1999). In 1933, Flosdorf and Mudd at the University of Pennsylvania performed lyophilization of blood serum under full aseptic conditions; they called that material lyophile (from the Greek λύος and φιλεῖν, solvent-loving) because of its ability to be rehydrated (Flosdorf and Mudd 1935).

The first industrial application of lyophilization dates back to World War II; this application was developed independently by Flosdorf and coworkers in the United States and by Greaves in England, for the production of dried plasma for wartime use and then for the production of penicillin (Flosdorf et al. 1945). The equipment used consisted of a single chamber for sublimation and condensation of vapor, vacuum pumps, and mechanical refrigeration apparatus. This plant configuration was used until the 1980s. Today, the vapor released by ice sublimation is generally condensed in a separate chamber.

Early commercial lyophilized products were Hemin by Abbott and Corticotrophin by Parker Davis and Rorer Pharmaceuticals. In the 1970s, new lyophilized products included antibiotics (penicillin G procaine by Wyeth Development, aminoglycosides by Wyeth, cephalosporins by Lilly, cefazolin by SKB and Lilly, β-lactams and vancomycin HCl by Lilly, tetracycline by Pfizer), vaccines (IBV H-52 and H-120 for infectious bronchitis), oncolytics (dactinomycin by Merck, cisplatin by BMS), two corticosteroids by Upjohn Company, hydrocortisone sodium succinate by Cortef, and methylprednisolone sodium succinate by Solu-Medrol.

In the 1980s and 1990s, new products were introduced for bacterial (Azactam by BMS, imipenem/cilastatin by Merck) and viral infections (acyclovir by GSK, ganciclovir by Syntex/Roche, interferons ra-2A by Roche, interferons ra-2B by Schering), treatment for multiple sclerosis (interferon rB1b by Bayer) and heart attack (alteplase by Genentech), and new vaccines (Cervarix by GSK for HPV prevention, Pentacel by Sanofi Pasteur for DPT + polio and hemophilus influenza B, Zostavax by Merck for herpes zoster) (Trappler 2013).

Nowadays, 40% of commercial biotherapeutics, including recombinant proteins, plasma, vaccines, and antibodies, and more than half of FDA-approved parenteral drugs require lyophilization (Akers 2010). Moreover, BCC Research reported that 16 of the top 100 (based on sales) pharmaceutical drugs are lyophilized (LaTorre-Snyder 2017). The single most lucrative product in 2017 was Enbrel (etanercept), a biologic product for autoimmune disease by Amgen and Pfizer, which generated global sales of $9 billion and that will lose its patent protection in 2019. Other examples of high-revenue products include the following: Remicade (infliximab) produced by Johnson and Johnson and Merck ($8 billion; patent expired in 2018); Herceptin (trastuzumab), a monoclonal antibody produced by Roche ($6.5 billion; patent expires in 2019); Copaxone (glatiramer acetate) manufactured by Teva Pharmaceuticals ($4.2 billion; patent expired in 2014).

In the last 20 years there has been a big change in how academia and industry approach research in the field. Kawasaki et al. (2019) described the evolution well, reporting that "the optimization study of the lyophilizer has been roughly developing by the order of (i) trial-and-error approach, (ii) process modeling using mathematical models, (iii) scalability, and (iv) quality-by-design."

The first breakthrough was the use of the mathematical approach to improve process understanding and to allow off-line optimization; the use of a mathematical model enabled the calculation of the "design space" with limited experimental effort, and to guide in safe and reliable process transfer and scale up. Process control also benefits from good modelling approaches, even if it has only been in the last years that the concepts of intelligent control have been applied to freeze-drying of pharmaceuticals.

The second breakthrough is process analytical technology (PAT), allowing the development of new PAT tools for monitoring the process, and innovations to allow control of the freezing step, enabling achievement of quality by design. According to Kawasaki et al. (2019) "a combination of PAT tools with a model/scale-up theory is expected to result in the QbD, i.e., a quality/risk management and an in-situ optimization of lyophilization operation. As important principles might be hidden behind the big data, for effective analysis, the use of the Internet of things (IoT) together with big data from PAT tool and the models including CFD would bring the rapid decision-making well fused with the practitioner's experiences." In addition, the use of advanced modelling tools, like molecular dynamics, has been recently proposed to better understand the complex interactions between active molecules and excipients or solid–liquid interfaces during both freezing and drying; these approaches offer a powerful tool for the "in silico" development of formulations to be lyophilized and reduce time and cost of experimentation.

The lyophilization equipment market is forecasted to double its value from $2.7 billion to $4.8 billion in 2020 as a consequence of development in the biopharma industry and the introduction of new drugs. The registration of new products with an expected high return, in particular, can be beneficial for introducing new technologies to production plants. Continuous freeze-drying might be one of these: this technology was developed for the coffee industry in the 1960s, but it has been object of recent interest by the pharmaceutic industry, as it has been shown that it can strongly reduce production time and is easily scalable, and enables quality control in each vial.

The new challenges, and the increasingly sophisticated approaches developed to respond to them, together with the new problems faced and solved by upcoming technologies, stimulated the production of new books or new editions.

This was the case for the second and third edition of Rey and May's book, *Freeze-Drying/Lyophilization of Pharmaceutical and Biological Products*, published in 2004 and 2010, respectively (Rey 2010). The editors of the present book contributed to that and various other books, including the five-volume series *Modern Drying Technology*, edited by Tsotsas and Mujumdar and completed in 2014, with chapters focusing on several of the previously mentioned breakthrough topics: the use of mathematical modeling and PAT for QbD, in-line product quality control, process monitoring, design space development, and equipment design (Barresi et al. 2010; Fissore and Barresi 2011; Fissore 2012, 2015; Fissore et al. 2015), management of non-uniform batches and scale-up (Pisano et al. 2011; Barresi and Pisano 2013), process intensification by means of combined technologies, use of organic solvents and control of nucleation (Pisano et al. 2014; Barresi et al. 2015; Pisano 2019), and advanced control (Barresi et al. 2018).

But in order to present organically the most recent advancements at the cutting edge of freeze-drying research and technology, from formulation design to process optimization and control, from new PAT monitoring tools and multivariate image analysis to process scale-down and development, from use of CFD for equipment design to development of continuous processes, the editors of the present volume were happy to accept Mujumdar's invitation to contribute to the series *Advances in Drying Science and Technology*, writing or co-writing several of the chapters and editing the book on *Freeze Drying of Pharmaceutical Products*. Thanks to the contribution of some lyophilization experts, this review work was carried out effectively.

Antonello Barresi

Roberto Pisano

Davide Fissore

REFERENCES

Akers, M. J. 2010. *Sterile drug products: Formulation, packaging, manufacturing and quality*. New York: CRC Press – Informa Healthcare.

Altmann, R. 1894. *Die Elementarorganismen und ihre Beziehungen zu den Zellen*. Liepzig: Veit.

Barresi, A. A., and R. Pisano. 2013. Freeze drying: Scale-up considerations. In *Encyclopedia of pharmaceutical science and technology*, 4th Edition, ed. J. Swarbrick, 1738–1752. New York: CRC Press (Taylor and Francis Group).

Barresi, A. A., Fissore, D., and D. Marchisio. 2010. Process analytical technology in industrial freeze-drying. In *Freeze-drying/lyophilization of pharmaceuticals and biological products*, 3rd rev. Edition, eds. L. Rey and J. C. May, Chap. 20, 463–496. New York: Informa Healthcare.

Barresi, A. A., Vanni, M., Fissore, D., and T. Zelenková. 2015. Synthesis and preservation of polymer nanoparticles for pharmaceuticals applications. In *Handbook of polymers for pharmaceutical technologies*. Vol. 2, *Processing and applications,* ed. V. K. Thakur and M. K. Thaku, Chap. 9, 229–280. Hoboken, NJ: John Wiley & Sons, Inc, and Salem, MA: Scrivener Publishing I.I.C.

Barresi, A. A., Pisano, R., and D. Fissore. 2018. Advanced control in freeze-drying. In *Intelligent control in drying*, ed. A. Martynenko and A. Bück, Chap. 19, 367–401. Boca Raton, FL: CRC Press (Taylor & Francis Group).

Fissore, D. 2012. Freeze drying. In *Encyclopedia of pharmaceutical science and technology*, 4th Edition, ed. J. Swarbrick, 1723–1737. New York: CRC Press (Taylor and Francis Group).

Fissore, D. 2015. Process analytical technology (PAT) in freeze drying. In *Advances in probiotic technology*, ed. P. Foerst and C. Santivarangkna, 264–285. Boca Raton, FL: CRC Press (Taylor & Francis Group).

Fissore, D., and A. A. Barresi. 2011. In-line product quality control of pharmaceuticals in freeze-drying processes. In *Modern Drying Technology Vol. 3: Product quality and formulation*, ed., E. Tsotsas and A. S. Mujumdar, Chap. 4, 91–154. Weinheim: Wiley-VCH Verlag GmbH & Co. KGaA.

Fissore, D., Pisano, R., and A. A. Barresi. 2015. Using mathematical modeling and prior knowledge for QbD in freeze-drying processes. In *Quality by design for biopharmaceutical drug product development,* ed. F. Jameel, S. Hershenson, M. A. Khan, and S. Martin-Moe, Chap. 23, 565–593. New York: Springer Science+Business Media.

Flosdorf, E. W., Hull, L. W., and S. Mudd. 1945. Drying by sublimation. *J. Immunol.* 50:21.

Flosdorf, E. W., and S. Mudd. 1935. Procedure and apparatus for preservation in "lyophile" form of serum and other biological substances. *J. Immunol.* 29:389–425.

Jennings, T. A. 1999. *Lyophilization: Introduction and basic principles*. Boca Raton, FL: Interpharm Press.

Kawasaki, H., Shimanouki, T., and Y. Kimura. 2019. Recent development of optimization of lyophilization process. *J. Chem.* 2019: Article ID 9502856, 14 pp.

LaTorre-Snyder, M. 2017. Lyophilization: The basics. An overview of the lyophilization process as well as the advantages and disadvantages. *Pharm. Process.* 32(1):1–2.

Pisano, R. 2019. Alternative methods of controlling nucleation in freeze-drying. In *Lyophilization of pharmaceuticals and biologicals: New technologies and approaches*, ed. K. R. Ward and P. Matejtschuk, Chap. 4, 74–111. New York: Springer Science+Business Media.

Pisano, R., Fissore, D., and A. A. Barresi. 2011. Heat transfer in freeze-drying apparatus. In *Developments in heat transfer*, ed. M. A. dos Santos Bernardes, Chap. 6, 91–114. Rijeka (Croatia).

Pisano, R., Fissore, D., and A. A. Barresi. 2014. Intensification of freeze-drying for the Pharmaceutical and food industry. In *Modern Drying Technology Vol. 5: Process intensification*, eds. E. Tsotsas and A. S. Mujumdar, Chap. 5, 131–161. Weinhein: Wiley-VCH Verlag GmbH & Co. KGaA.

Trappler, E. H. 2013. Contemporary approaches to development and manufacturing of lyophilized parenterals. In *Sterile product development*, ed. P. Kolhe, M. Shah and N. Rathore, 275–313. New York: Springer.

Rey, L. 2010. Glimpses into the realm of freeze-drying: Classical issues and new ventures. In *Freeze-drying/lyophilization of pharmaceutical and biological products*, ed. L. Rey and J. C. May, 1–32. New York: CRC Press.

Editors

Davide Fissore is Professor of Chemical Engineering at Politecnico di Torino (Italy). His research activity is mainly focused on process modelling and optimisation, and on the design and validation of model-based tools for process monitoring and control. One of the topics of his research activity is the freeze-drying of pharmaceutical products and foodstuffs. He developed various devices to monitor and optimise the in-line (using a control system) or off-line (using the design space of the product) freeze-drying process for a given product. He acted as a consultant for several pharmaceutical companies, focusing on process development and scale up. Davide Fissore is author or co-author of 90 papers appeared in international peer-reviewed journals and 15 book chapters, and he currently holds 9 patents, issued or pending.

Roberto Pisano is a Professor of Chemical Engineering at Politecnico di Torino (Italy), where he earned a PhD in 2009. Professor Pisano's research focuses on the application of both computational and experimental methods to engineering chemical products and processes, with particular emphasis on pharmaceutical processing and formulation of both small molecules and biologics. He was a visitor researcher at Centre de Ressources Technologiques—Institut Technique Agro-Industriel (Strasbourg, France) in 2008 and at the Department of Chemical Engineering of Massachusetts Institute of Technology (Cambridge, USA) in 2016. He has worked with many pharmaceutical companies in research or consulting. He has published more than 75 papers and 7 book chapters and currently has 4 patents issued or pending.

Antonello Barresi is currently full Professor of Transport Phenomena at Politecnico di Torino (Italy), in charge of the course on Process Development and Design. Currently he serves as Italian national delegate to the Working Party on Drying for the European Federation of Chemical Engineers. His main research interests in drying include drying and freeze-drying of pharmaceuticals and enzymes, modelling and optimization of freeze-drying processes, and control of industrial freeze-dryers. Most recent research is focused on process transfer, scale-up and cycle development, and new approaches for process development and quality control in freeze-drying of pharmaceutical and food products. He is the author of more than 250 papers (of which about 160 are published in international journals or books) and more than 100 conference presentations.

Contributors

Andrea Arsiccio
Dipartimento di Scienza Applicata e
  Tecnologia
Politecnico di Torino
Torino, Italy

Antonello Barresi
Dipartimento di Scienza Applicata e
  Tecnologia
Politecnico di Torino
Torino, Italy

Luigi C. Capozzi
Dipartimento di Scienza Applicata e
  Tecnologia
Politecnico di Torino
Torino, Italy

Domenico Colucci
Dipartimento di Scienza Applicata e
  Tecnologia
Politecnico di Torino
Torino, Italy

Jos A. W. M. Corver
RheaVita
Ghent, Belgium

Alberto Ferrer
Department of Applied Statistics
  and Operational Research and
  Quality
Universitat Politècnica de València
Valencia, Spain

Davide Fissore
Dipartimento di Scienza Applicata e
  Tecnologia
Politecnico di Torino
Torino, Italy

Wolfgang Friess
Department of Pharmacy,
  Pharmaceutical Technology and
  Biopharmaceutics
Ludwig-Maximilians-Universität
Munich, Germany

Yowwares Jeeraruangrattana
Health and Life Sciences Faculty
Leicester School of Pharmacy
De Montfort University
Leicester, United Kingdom

Roberto Pisano
Dipartimento di Scienza Applicata e
  Tecnologia
Politecnico di Torino
Torino, Italy

José Manuel Prats-Montalbán
Department of Applied Statistics and
  Operational Research and Quality
Universitat Politècnica de València
Valencia, Spain

Ivonne Seifert
Department of Pharmacy
Pharmaceutical Technology and
  Biopharmaceutics
Ludwig-Maximilians-Universität
Munich, Germany

Geoff Smith
Health and Life Sciences Faculty
Leicester School of Pharmacy
De Montfort University
Leicester, United Kingdom

Taylor N. Thompson
Millrock Technology, Inc.
Kingston, New York

1 The Freeze-Drying of Pharmaceutical Products

Introduction and Basic Concepts

*Davide Fissore, Roberto Pisano,
and Antonello Barresi*

CONTENTS

1.1 INTRODUCTION

Freeze-drying (or lyophilization) is a key step in the production process of several pharmaceutical products as it allows recovering an active pharmaceutical ingredient from a liquid solution at low temperature; this increases the shelf-life of the product and preserves critical product quality attributes because the process is carried out at low temperature and the drugs are generally thermolabile molecules (Jennings 1999; Mellor 2004; Oetjen and Haseley 2004; Fissore 2013).

In a freeze-drying process the product temperature is lowered to a value well below the water freezing point. In this way most of the water, the so-called "free water," turns into the frozen stage (see Figure 1.1), while part of the water remains bound to the product molecules.

Then, chamber pressure is lowered to a value well below the water triple point: in this way the ice may sublimate (primary drying) if heat is supplied to the product, because sublimation of the ice is an endothermal process. Finally, the target value of residual moisture in the final product is obtained by removing (most of) the bound

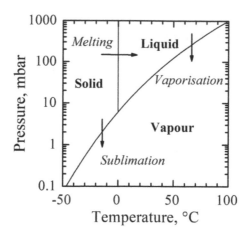

FIGURE 1.1 Water phase diagram.

water (secondary drying). In this stage product temperature is further increased with respect to the value reached during primary drying, as chamber pressure may be reduced, although some studies evidenced no improvements when secondary drying is carried out at a very low pressure (Pikal et al. 1980). The freeze-drying equipment, the objectives and the constraints of the process, the role of mathematical modelling and the open issues are briefly presented and discussed in the following sections.

1.2   FREEZE-DRYING EQUIPMENT

Figure 1.2 shows a sketch of a batch freeze-dryer, representing the current standard equipment used to carry out the freeze-drying of a pharmaceutical product (Oetjen and Haseley 2004).

The liquid solution containing the active pharmaceutical ingredient and the excipients (buffers, bulking agents, etc.) is poured into glass vials that may be loaded directly onto the shelves of the drying chamber or through trays that are placed onto the shelves. As both the ice sublimation (primary drying) and water desorption (secondary drying) are endothermic processes, it is necessary to supply heat to the product, and this is done through a hot fluid flowing inside the shelves. Therefore, the use of a tray, although it can make the loading/unloading of the vials easier, increases the resistance to heat transfer from the heating fluid, which results in a lengthening of the drying process. As an alternative, bulk freeze-drying may be carried out by filling a tray with the liquid solution; in this case at the end of the process the powder has to be collected and further processed to prepare the unit doses.

Before being loaded in the drying chamber the liquid product may be frozen in separate equipment, although the standard approach is to freeze the liquid solution directly in the drying chamber. To this purpose, the temperature of the fluid flowing inside the shelves of the freeze-dryer is lowered, down to −40/−50°C, in such a way that the "free" water may turn into the frozen state. In some cases, controlled nucleation techniques may be used, aiming at inducing the nucleation of the ice crystals in

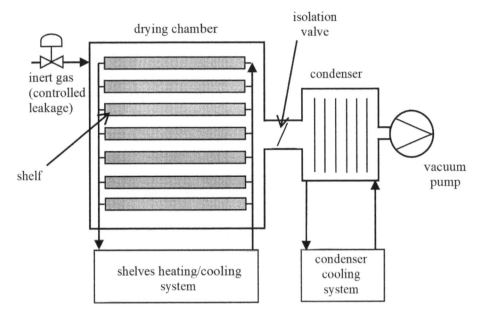

FIGURE 1.2   Scheme of a batch freeze-dryer.

the vials at the same time (and temperature), thus obtaining a more uniform ice crystal size (Searles et al. 2001). In fact, after sublimation, the size of the empty spaces in the dried product corresponds to the size of the ice crystals, and a uniform structure of the dried product allows reduction of the intra-vial variability. Several techniques are currently available to induce nucleation at the same time in all the vials of the batch (and will be discussed in greater detail in Chapter 6):

1. The "ice fog" technique: ice crystals are formed in the drying chamber when a flow of nitrogen at very low temperature is introduced in the humid drying chamber (Patel et al. 2009). As an alternative, the ice fog may be obtained in the condenser where a certain amount of water is atomized, provided that the temperature of the condenser has been lowered (Ling 2011; Thompson 2013; Wollrath 2019). Once the ice crystals are obtained, they enter the vials, causing nucleation. The main concern of this technique is related to obtaining a uniform distribution of the ice crystals in the chamber, in such a way that ice nucleation occurs effectively at the same time in all the vials.
2. Ultrasonic vibrations (at frequencies above 10 kHz): they can cause the formation of small-size gas bubbles that rapidly grow and collapse. This causes very high pressure and temperature fluctuations that induce ice nucleation (Morris et al. 2004; Nakagawa et al. 2006)
3. Pressurization/depressurization: Konstantinidis et al. (2011) proposed a method based on the creation of a pressure fluctuation as a method to induce ice nucleation. The chamber of the freeze-dryer, after vials loading,

is pressurized with argon to about 2.80–2.95 bar (26–28 psig) at the beginning of the run. Then, when the target nucleation temperature is reached,
the pressure is rapidly reduced to almost atmospheric pressure (about 1
psig) to induce nucleation.

4. Vacuum-induced surface freezing: when the desired nucleation temperature is reached, chamber pressure is reduced for a short time interval. This
   causes water evaporation at the top of the product and causes ice nucleation
   (Kramer et al. 2002; Liu et al. 2005; Oddone et al. 2014).

Water vapor produced in the drying chamber flows to the condenser, where the low
temperature of the condenser surface promotes vapor frosting. A critical issue is
represented by size of the duct connecting the drying chamber to the condenser, as
sonic flow may be reached, resulting in a loss of pressure control in the chamber.

With respect to pressure control in the drying chamber, a vacuum pump, connected to the condenser, is generally used, although in some cases a flow of nitrogen,
or other inert gas, in the drying chamber (controlled leakage) may be used to get a
tighter control of chamber pressure.

1.3   PROCESS OBJECTIVES AND CONSTRAINTS

Although it is considered a "gentle" process toward the product, the freeze-drying
process may compromise the quality of the final product. One of the main constraints that must be respected is the temperature of the product. Each formulation
being processed is in fact characterized by a threshold temperature that should not
be trespassed during the process. In the case of amorphous products, this limit value
is related to the temperature causing the collapse of the dried product (Pikal 1994).
Collapse results in a blockage of the pores, thus increasing the resistance of the
dried product to vapor flow and increasing the duration of the process. Moreover, it
may be responsible for a higher residual amount of water in the product at the end
of the process, for a higher reconstitution time, and, in some cases, for the activity
loss of the drug (Bellows and King 1972; Tsourouflis et al. 1976; Adams and Irons
1993; Wang et al. 2004). Finally, a collapsed product is generally rejected owing to
the unattractive physical appearance. In the case of crystalline products, the limit
temperature corresponds to the eutectic temperature, to avoid the formation of a
liquid phase.

When dealing with product temperature one must bear in mind that the vials of
a batch do not experience the same dynamics throughput the drying process. This is
due to several reasons, such as non-uniform heating conditions in the drying stages
(for example, due to radiation from chamber walls) and pressure gradients in the drying chamber (Rasetto et al. 2010). In addition, the freezing stage, when no controlled
nucleation is used, also may be a cause of non-uniformity in the drying stages, as the
non-uniform inter-vial cake structure is responsible for a non-uniform sublimation
flux among the vials.

The main goal of the process is thus to keep product temperature below the limit
value in all the vials of the batch. In addition to this, the duration of the process must
be minimized, as the freeze-drying process of a given product may be a very long
process over time.

In addition to product temperature, the equipment is also a constraint. The condenser capacity has to be compatible with the sublimation rate, and, above all, the water vapor flux from the chamber has to be lower than the value for which there is sonic flow in the duct (Searles 2004; Patel et al. 2010; Fissore et al. 2015; Marchisio et al. 2018).

1.4 MATHEMATICAL MODELLING AND PROCESS DESIGN

To meet the goals and the constraints previously listed, the operating conditions of the drying process, namely the values of the temperature of the fluid flowing inside the shelves (or, what is the same, the temperature of the heating shelf) and of the pressure in the drying chamber have to be carefully selected. An extended experimental campaign is generally required to assess the effect of the operating conditions on product temperature and sublimation flux (and, ultimately, of the drying duration). As an alternative, to save time, mathematical modelling may be useful, as it allows simulating *in silico* the primary drying of a given product; experiments are also required in this case, to get the values of model parameters and for model validation, but the experimental effort is surely lower (Kawasaki et al. 2019).

In the stage of process development, it is not only required to identify the most suitable operating conditions but also to assess the robustness of the selected cycle. The design space is the tool that may be used for this purpose. The "ICH Q8 Pharmaceutical Development Guideline" (2009) defines the design as the multidimensional combination of input variables and process parameters that have been demonstrated to provide assurance of quality. In case the design space is known it is possible to evaluate the safety margins of the selected operating conditions, that is, how far the temperature of the product is from the limit value.

In the past, detailed multi-dimensional models were proposed for both the primary and the secondary drying stages (Liapis and Bruttini 1995). Unfortunately, the calculations required by these models are highly time-consuming, and their equations involve several parameters whose values are quite often unknown and/or could be estimated only with high uncertainty. This can jeopardize the higher accuracy of this type of model. As an alternative, mono-dimensional models may be used: they assume that temperature and composition gradients in the product are negligible in the radial direction (as confirmed, experimentally, by Pikal 1985), and thus, a flat interface, separating the frozen and the dried product, is established in the primary drying stage, which moves from the top to the bottom of the product in the vial (Figure 1.3).

One of the most frequently used mono-dimensional models used for process simulation is constituted by the following equations (Velardi and Barresi 2008):

$$\frac{dL_f}{dt} = -\frac{1}{\rho_f - \rho_d} J_w, \tag{1.1}$$

$$J_q = \Delta H_s J_w, \tag{1.2}$$

$$T_B = T_s - \frac{1}{K_v} \left(\frac{1}{K_v} + \frac{L_f}{k_f} \right)^{-1} (T_s - T_i), \tag{1.3}$$

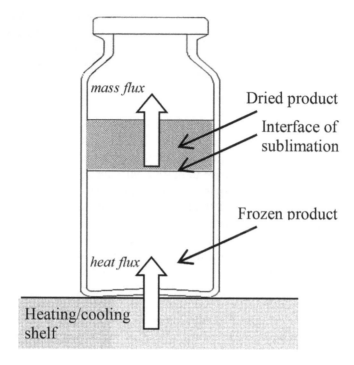

FIGURE 1.3   Schematic of the primary drying stage of a freeze-drying process in vials.

$$J_q = K_v \left(T_s - T_B \right),$$ (1.4)

$$J_w = \frac{1}{R_p} \left(p_{w,i} - p_{w,c} \right).$$ (1.5)

In previous equations J_q is the heat flux to the product, J_w is the mass flux from the interface of sublimation to the drying chamber, K_v is the heat transfer coefficient, R_p is the resistance of the dried product to vapor flux, ρ_f and ρ_d are, respectively, the density of the frozen and of the dried product, ΔH_s is the heat of sublimation, k_f is the thermal conductivity of the frozen product, L_f is the thickness of the frozen product, t is the time, T_i and T_B are, respectively, the temperature of the interface of sublimation and that of the product at the bottom of the vial, $p_{w,i}$ and $p_{w,c}$ are, respectively, the water vapor partial pressure at the interface of sublimation, which is a known function of T_i (Fissore et al. 2011a), and in the drying chamber (which is generally assumed to coincide with the total chamber pressure, because it is the chamber atmosphere with about 100% water vapor). Equation 1.1 is the mass balance of the frozen product; equation 1.2 is the energy balance at the interface of sublimation, and equation 1.3 is obtained from the energy balance for the frozen product.

This simple model was used by Giordano et al. (2010) and by Fissore et al. (2011b) for calculating the design space of the primary drying stage of the freeze-drying process, as this is the longest stage of the whole process and the point at which the limit temperature is lower, thus motivating the modeling effort. A simpler approach is that of Koganti et al. (2011), also in this case based on a one-dimensional approach.

In order to use even a simple model like that expressed by eqs. (1.1)–(1.5) it is necessary to know the value of the parameters K_v and R_p. Several techniques were proposed in the past, based on the available process analytical technologies, recently reviewed by Fissore et al. (2018).

1.5 OPEN ISSUES

Several issues are still under discussion in the field of freeze-drying of pharmaceutical products, which constitute the topics of the following chapters. A brief summary is given in the following.

1.5.1 FORMULATION DESIGN

A trial-and-error approach is generally used to find out the best excipients, aiming to increase the threshold temperature of the product and to best preserve the pharmaceutical activity of the drug at the end of the process. Previous experiences in the field may help to choose the excipients, but this does not guarantee that the optimal result is achieved, and, in any case, several experiments are needed (and, quite often, they are not compatible with the time to market of new drugs). Chapter 2 addresses the use of molecular dynamics for formulation design and optimization, a powerful tool that allows the *in silico* simulation of the formulation behavior during freezing and drying stages, while Chapter 3 is focused on the novel excipients that are the subject of several research studies aimed at identifying candidates to replace the well-established sugars as freeze-drying excipients.

1.5.2 PROCESS MONITORING AND CONTROL

Several process analytical technologies were proposed in the last decade for monitoring the freeze-drying process, aiming to measure/estimate the temperature of the product, the residual amount of ice, the sublimation flux, and the values of the parameters of the mathematical model to be used for *in silico* process simulation (see the recent review by Fissore et al. 2018). Recent innovations deal with the use of multivariate image analysis (Chapter 4) and impedance methods (Chapter 5) for process monitoring and with the use of model-based control systems for in-line process optimization (Chapter 6).

1.5.3 SMALL-SCALE FREEZE-DRYERS

The process development phase is time-consuming as several tests are needed to assess the effect of the operating conditions on drying duration and critical product quality attributes and because each test takes a long time owing to batch preparation, loading/unloading, condenser defrosting, and so on. Additionally, the consumption of the drug may be a concern, as it can be expensive or even unavailable in a sufficient amount in this stage. It is therefore particularly important to have the possibility of using micro freeze-dryers, where just few vials are processed, for process development: they allow saving raw materials, beside the time related to batch preparation and testing (Chapter 7).

1.5.4 CONTINUOUS FREEZE-DRYING

The need for faster processes and improved quality control is the driving force for the development of new freeze-drying concepts. The possibility of carrying out the freeze-drying of pharmaceutical products in a continuous mode, as for food products, is surely the most significant innovation in the field. This is the topic of Chapter 8.

1.5.5 EQUIPMENT DESIGN

Despite the importance of freeze-drying equipment on the process and the quality of the product, this topic has not been deeply investigated. Computational fluid dynamics stands out as a powerful tool for addressing these issues, aiming to improve the design of the drying chamber, the condenser, and the valve in the duct connecting the chamber to the condenser in such a way that the limit to the sublimation flux posed by the equipment capacity is higher and the drying conditions in the chamber are more uniform (Chapter 9).

REFERENCES

Adams, G. D. J., and L. I. Irons. 1993. Some implications of structural collapse during freeze-drying using Erwinia caratovora Lasparaginase as a model. *J. Chem. Technol. Biotechnol.* 58:71–76.

Bellows, R. J., and C. J. King. 1972. Freeze-drying of aqueous solutions: Maximum allowable operating temperature. *Cryobiology* 9:559–561.

Fissore, D. 2013. Freeze-drying of pharmaceuticals. In *Encyclopedia of pharmaceutical science and technology,* 4th Edition, ed. J. Swarbrick, 1723–1737. London: CRC Press.

Fissore, D., Pisano, R., and A. A. Barresi. 2011a. On the methods based on the Pressure Rise Test for monitoring a freeze-drying process. *Drying Technol.* 29:73–90.

Fissore, D., Pisano, R., and A. A. Barresi. 2011b. Advanced approach to build the design space for the primary drying of a pharmaceutical freeze-drying process. *J. Pharm. Sci.* 100:4922–4933.

Fissore, D., Pisano, R., and A. A. Barresi. 2015. Using mathematical modeling and prior knowledge for QbD in freeze-drying processes. In *Quality by design for biopharmaceutical drug product development*, ed. F. Jameel, S. Hershenson, M. A. Khan, and S. Martin-Moe, 565–593. New York: Springer.

Fissore, D., Pisano, R., and A. A. Barresi. 2018. Process analytical technology for monitoring pharmaceuticals freeze-drying – A comprehensive review. *Drying Technol.* 36:1839–1865.

Giordano, A., Barresi, A. A., and D. Fissore. 2010. On the use of mathematical models to build the design space for the primary drying phase of a pharmaceutical lyophilization process. *J. Pharm. Sci.* 100:311–324.

International Conference on Harmonisation of Technical requirements for Registration of Pharmaceuticals for Human Use. 2009 ICH Harmonised Tripartite Guideline. Pharmaceutical Development Q8 (R2).

Jennings, T. A. 1999. *Lyophilization: Introduction and basic principles.* Boca Raton, FL: Interpharm/CRC Press.

Kawasaki, H., Shimanouki, T., and Y. Kimura. 2019. Recent development of optimization of lyophilization process. *J. Chem.* 2019: Article ID 9502856, 14 pp.

Koganti, V. R., Shalaev, E. Y., Berry, M. R., et al. 2011. Investigation of design space for freeze-drying: Use of modeling for primary drying segment of a freeze-drying cycle, *AAPS PharmSciTech.* 12:854–861.

Konstantinidis, A. K., Kuu, W., Otten, L., et al. 2011. Controlled nucleation in freeze-drying: Effects on pore size in the dried product layer, mass transfer resistance, and primary drying rate. *J. Pharm. Sci.* 100:3453–3470.

Kramer, M., Sennhenn, B., and G. Lee. 2002. Freeze-drying using vacuum-induced surface freezing. *J. Pharm. Sci.* 91:433–443.

Liapis, A. I., and R. Bruttini. 1995. Freeze-drying of pharmaceutical crystalline and amorphous solutes in vials: Dynamic multi-dimensional models of the primary and secondary drying stages and qualitative features of the moving interface. *Drying Technol.* 13:43–72.

Ling, W. 2011. Controlled nucleation during freezing step of freeze-drying cycle using pressure differential ice fog distribution. U.S. Patent Application 2012/0272544 A1 filed April 29, 2011.

Liu, J., Viverette, T., Virgin, M., et al. 2005. A study of the impact of freezing on the lyophilization of a concentrated formulation with a high fill depth. *Pharm. Dev. Technol.* 10:261–272.

Marchisio, D. L., Galan, M., and A. A. Barresi. 2018. Use of Computational Fluid Dynamics for improving freeze-dryers design and process understanding. Part 2: Condenser duct and valve modelling. *Eur. J. Pharm. Biopharm.* 129:45–57.

Mellor, J. D. 2004. *Fundamentals of freeze-drying.* London: Academic Press.

Morris, J., Morris, G. J., Taylor, R., et al. 2004. The effect of controlled nucleation on ice structure, drying rate and protein recovery in vials in a modified freeze dryer. *Cryobiology* 49:308–309.

Nakagawa, K., Vessot, S., Hottot, A., and J. Andrieu. 2006. Influence of controlled nucleation by ultrasounds one ice morphology of frozen formulations for pharmaceutical proteins freeze-drying. *Chem. Eng. Proc.* 45:783–791.

Oddone, I., Pisano, R., Bullich, R., and P. Stewart. 2014. Vacuum-induced nucleation as a method for freeze-drying cycle optimization. *Ind. Eng. Chem. Res.* 53:18236–18244.

Oetjen, G. W., and P. Haseley. 2004. *Freeze-drying.* Weinheim: Wiley-VHC.

Patel, S. M., Bhugra, C., and M. J. Pikal. 2009. Reduced pressure ice fog technique for controlled ice nucleation during freeze-drying. *AAPS PharmSciTech.* 10:1406–1411.

Patel, S. M., Swetaprovo, C., and M. J. Pikal. 2010. Choked flow and importance of Mach I in freeze-drying process design. *Chem. Eng. Sci.* 65:5716–5727.

Pikal, M. J. 1985. Use of laboratory data in freeze-drying process design: Heat and mass transfer coefficients and the computer simulation of freeze-drying. *J. Parenter. Sci. Technol.* 39:115–139.

Pikal, M. J. 1994. Freeze-drying of proteins: Process, formulation, and stability. In *Formulation and delivery of proteins and peptides*, ed. J. L. Cleland and R. Langer, 12–33. Washington, DC: American Chemical Society.

Pikal, M. J., Shah, S., Roy, M. L., and R. Putman. 1980. The secondary drying stage of freeze drying: Drying kinetics as a function of temperature and pressure. *Int. J. Pharm.* 60:203–217.

Rasetto, V., Marchisio, D. L., Fissore, D., and A. A. Barresi. 2010. On the use of a dual-scale model to improve understanding of a pharmaceutical freeze-drying process. *J. Pharm. Sci.* 99:4337–4350.

Searles, J. 2004. Observation and implications of sonic water vapour flow during freeze-drying. *Am. Pharm. Rev.* 7:58–69.

Searles, J., Carpenter, J., and T. Randolph. 2001. The ice nucleation temperature determines the primary drying rate of lyophilization for samples frozen on a temperature-controlled shelf. *J. Pharm. Sci.* 90:860–871.

Thompson, T. N. 2013. LyoPAT™: Real-time monitoring and control of the freezing and primary drying stages during freeze-drying for improved product quality and reduced cycle times. *Am. Pharm. Rev.* 16:68–74.

Tsourouflis, S., Flink, J. M., and M. Karel. 1976. Loss of structure in freeze-dried carbohydrates solutions: Effect of temperature, moisture content and composition. *J. Sci. Food Agric.* 27:509–519.

Velardi, S. A., and A. A. Barresi. 2008. Development of simplified models for the freeze-drying process and investigation of the optimal operating conditions. *Chem. Eng. Res. Des.* 86:9–22.

Wang, D. Q., Hey, J. M., and S. L. Nail. 2004. Effect of collapse on the stability of freeze-dried recombinant factor VIII and a-amylase. *J. Pharm. Sci.* 93:1253–1263.

Wollrath, I., Friess, W., Freitag, A., Hawe, A., and G. Winter. 2019. Comparison of ice fog methods and monitoring of controlled nucleation success after freeze-drying. *Int. J. Pharm.* 558:18–28.

2 Formulation Design and Optimization Using Molecular Dynamics

Roberto Pisano and Andrea Arsiccio

CONTENTS

2.1 DENATURATION STRESSES DURING LYOPHILIZATION

The global biopharmaceuticals market has grown considerably in the last few years and is projected to exhibit a remarkable expansion in the near future. In fact, protein-based drugs are playing an increasing role in the treatment of a wide number of human diseases, including diabetes, autoimmune diseases, and numerous forms of cancer.

The most common technique for preparing solid protein-based therapeutics is freeze-drying, or lyophilization, which allows removal of water at low temperature. Despite being a mild process, freeze-drying can nevertheless result in undesired disruption of the protein tertiary and secondary structure or in the formation of aggregates (Wang and Roberts 2010; Roberts 2014; Wang 2000).

This can result in a reduction of therapeutic activity, and in some cases it may also lead to serious safety issues, such as generation of an immune response (Roberts 2014; Jiskoot et al. 2012).

The freeze-drying stresses are related to both the freezing and the drying steps. For instance, the interaction of nonpolar residues with water is less unfavorable at the low temperature used during freezing, and this may lead to cold denaturation of the protein (Privalov 1990; Franks 1995; Graziano et al. 1997; Lopez et al. 2008; Matysiak et al. 2012). The formation of ice may also be harmful to protein stability, for several reasons. First, proteins may adsorb to the interface, and undergo surface-induced denaturation (Strambini and Gabellieri 1996). Since a high cooling rate results in small ice crystals, with a large surface area, it has been often observed that protein unfolding is enhanced at higher cooling rates (Chang et al. 1996). In addition to this, the crystallization of formulation components, such as glycine and mannitol, may also result in the formation of an additional interface, where surface-induced denaturation may occur (Liu et al. 2005; Al-Hussein and Gieseler 2012). Another consequence of ice formation is the rapid increase in solutes concentration, which combined to selective crystallization of formulation components may lead to significant changes in ionic strength and relative composition of the amorphous phase, possibly destabilizing a protein (Al-Hussein and Gieseler 2012). As a result of the increased solute concentration, chemical reactions may actually accelerate in a partially frozen aqueous solution (Pikal 2010).

The effects for protein stability may be particularly significant when the component undergoing selective crystallization is a buffering species, as this could cause important pH changes. This may happen, for instance, with sodium phosphate and potassium phosphate (van den Berg and Rose 1959; Anchordoquy and Carpenter 1996; Gómez et al. 2001; Murase and Franks 1989).

The presence of polymers in the formulation may also cause problems of phase separation, owing to the polymers altered solubility at low temperature (Heller et al. 1997). Phase separation leads to the formation of an additional interface where the protein may adsorb and denature and creates shear stresses that could be harmful to protein stability. Furthermore, the protein may preferentially partition into a phase with very low concentration of the stabilizer, reducing the protective effect.

Finally, the freeze-drying process removes part of the hydration shell of the protein, which may disrupt the native state of the protein (Rupley and Careri 1991). This is due to both a decreased charge density on the protein surface in a water-poor environment, which promotes aggregation, and the potential removal of water molecules that are an integral part of active sites in the protein.

Because of the risks of protein denaturation associated to the freeze-drying process, it is of utmost importance to design a formulation capable of minimizing the loss of therapeutic activity. In this context, we will describe in the following sections the possible mechanisms of protein stabilization during both the freezing and drying steps of lyophilization, and some examples of common pharmaceutical excipients will be provided. Afterwards, the current approach for formulation design, based on design of experiments (DoE) and on practical considerations, will be introduced briefly. Finally, we will show how the *in silico* modeling technique known as molecular dynamics (MD) may help optimize the formulation and guide the selection of excipients from a knowledge-based point of view. The main features of MD simulations will be shortly described, and examples of applications to the design of freeze-dried formulations will be provided.

2.2 MECHANISMS OF PROTEIN STABILIZATION DURING FREEZE-DRYING

The most common excipients used for protein stabilization during freeze-drying include

- sugars, such as the disaccharides sucrose, trehalose, and lactose or the monosaccharide glucose
- polyols, such as sorbitol and glycerol
- polymers, including albumin, dextran, polyvinylpyrrolidone (PVP) or hydroxyethyl cellulose (HEC)
- amino acids, such as glycine, proline, arginine, etc.
- surfactants, especially the polysorbates Tween 20 and Tween 80

When designing a freeze-dried formulation, it is important to remember that freezing and drying expose proteins to different stresses; therefore, the mechanisms of protein stabilization by excipients are not the same during the two stages of lyophilization. As a general guideline, those excipients that stabilize a protein in solution also have a protective action during freezing, as in both cases water is present. However, the mechanisms of protein stabilization are different in the dried state, and in this case the ability of the excipients to form a stiff, compact cake that inhibits the protein motions responsible for unfolding and aggregation becomes dominant (Ohtake et al. 2011). In the following, the main mechanisms of cryo- and lyoprotection will be described. The role of surfactants will be discussed in a separate section, as their mechanism is significantly different.

2.2.1 MECHANISMS OF CRYOPROTECTION

One of the most widely accepted mechanisms of protein stabilization during freezing is preferential exclusion. According to this theory, a protective osmolyte should be excluded from the protein surface, and preferential hydration should therefore ensue.

The presence of these cosolutes creates a thermodynamically unfavorable situation, which is augmented by an increase in the surface area exposed by the protein (Timasheff 1993).

As a consequence, the native fold is stabilized, because denaturation would result in a greater contact area between the protein and the solvent. Besides, stabilization of the native structure is greater in case of high excipient concentrations because preferential exclusion increases with the concentration of the osmolyte (Arakawa and Timasheff 1982).

The preferential exclusion of an excipient from the protein surface can be quantified using the Kirkwood-Buff integrals, G_{ij} (Kirkwood and Buff 1951; Ben-Naim 2006),

$$G_{ij} = \int_{r=0}^{\infty} \left[g_{ij}(r) - 1 \right] 4\pi r^2 dr, \tag{2.1}$$

where $g_{ij}(r)$ is the radial distribution function for component i with respect to component j at a given distance r between i and j. With the Kirkwood-Buff integrals it is possible to compute the preferential exclusion coefficient Γ_{23} between the protein (2) and the excipient (3), as

$$\Gamma_{23} = \rho_3 (G_{23} - G_{21}), \tag{2.2}$$

where G_{21} and G_{23} are the protein-water (1) and protein-excipient Kirkwood-Buff integrals, respectively, while ρ_3 is the number density of the excipient in the solution. A positive value of the preferential exclusion coefficient indicates preferential interaction between the protein and the excipient, while a negative Γ_{23} implies preferential exclusion.

Preferentially excluded excipients also hinder dissociation of proteins with a quaternary structure, again because the dissociation process would increase the surface area and the thermodynamic instability. However, those osmolytes that are excluded from contact with the protein may also enhance aggregation phenomena, because the self-association of proteins reduces the total surface area exposed. This undesired intensification of self-association is anyway prevented if protein unfolding is the key determinant in causing aggregation; in this case the excluded cosolute should be able to reduce aggregation by stabilizing the native structure (Ohtake et al. 2011).

In addition to this thermodynamic mechanism, many common cryoprotectants, such as polymers and sugars, also increase the viscosity of the solution, especially as a result of cryoconcentration. The high viscosity that is eventually reached dramatically restricts diffusion processes and should consequently decrease the protein unfolding rate. Furthermore, some cryoprotectants may stabilize proteins by suppressing pH changes during freezing (Anchordoquy and Carpenter 1996).

Some other hypotheses have been proposed more recently. For instance, it was found that the favorable interaction, rather than preferential exclusion, between lactate dehydrogenase (LDH) and high-molecular-weight polyethylene glycols (PEGs) (e.g., PEG 4000 and PEG 8000) cryoprotected LDH (Mi et al. 2004). However, preferential exclusion remains by far the most widely used theory to explain protein stabilization both in solution and during freezing.

2.2.2 MECHANISMS OF LYOPROTECTION

During the drying steps of freeze-drying, the preferential exclusion mechanism is no longer applicable because at this point there is essentially no water (Carpenter et al. 1993). Therefore, different theories have been developed to explain protein stability in the dried state. For instance, according to the water replacement mechanism (Crowe et al. 1984; Carpenter and Crowe 1989; Carpenter et al. 1990), protective osmolytes should hydrogen bond with the protein at the end of the drying process to satisfy the hydrogen-bonding requirement of the polar residues. In fact, a protein may form nonnative intermolecular hydrogen bonds upon dehydration, resulting in unfolding. Formation of the intermolecular hydrogen bonds is inhibited if the excipients serve as water substitutes.

However, since the highest number of hydrogen bonds can be formed if the stabilizer is in the amorphous phase, crystallization of the excipient may reduce protein stabilization (Carpenter et al. 1993).

One other major mechanism of protein stabilization by lyoprotectants is the formation of a viscous, glassy matrix during lyophilization (Green and Angell 1989; Franks 1994) that increases protein stability by slowing down protein denaturation and unfolding (Hagen et al. 1995). Thus, in this case stabilization occurs via a pure kinetic mechanism, and a requirement is that the protein and stabilizer are in the same amorphous phase.

Finally, according to the water entrapment hypothesis, excipients may form a cage around the protein that entraps and slows down water molecules (Belton and Gil 1994; Corradini et al. 2013). This makes it possible both to maintain a high level of hydration and to hinder the protein motions that may lead to denaturation.

2.2.3 ROLE OF SURFACTANTS

As previously discussed, a major cause of biopharmaceuticals' instability is related to phenomena of surface-induced denaturation and aggregation. Therefore, surfactants are often added to the formulation in order to prevent surface-driven damage. The polysorbates Tween 20 and Tween 80, which are nonionic surfactants consisting of fatty acid esters of polyethylene sorbitan, are commonly used for protein stabilization.

Different mechanisms have been proposed to explain the protective action of surfactants (Lee et al. 2011; Randolph and Jones 2002). For instance, surfactants preferentially locate at interfaces, thus reducing adsorption and/or aggregation phenomena. This effect is concentration-dependent and correlates with the critical micelle concentration (CMC) of the surfactant.

It was also suggested that surfactants may assist in protein refolding, acting like molecular chaperones, and should increase the free energy of protein unfolding (Bam et al. 1996).

Finally, it is also possible that surfactants associate with proteins in solution, preventing self-association. This last effect should not be related to the CMC of the surfactant, but to the molar binding stoichiometry between the surfactant and the protein (Bam et al. 1995, 1998; Deechongkit et al. 2009).

However, numerous unresolved questions remain, together with some contradictory observations. For instance, Deechongkit et al. (Deechongkit et al. 2009) observed that polysorbates affected the α-helix content of darbepoetin alpha, even below the CMC. In contrast, previous studies of recombinant human growth hormone (rhGH) (Bam et al. 1996) and anti-L-selectin antibody (Jones et al. 2001) by far-UV circular dichroism seemed to demonstrate a negligible effect of surfactant molecules on the proteins' secondary structure.

2.3 CURRENT APPROACH TO FORMULATION DESIGN

The challenge for a formulation scientist is to determine the optimal choice of excipients and the best concentration of each excipient. Currently, the design and optimization of pharmaceutical formulations are mainly based on some practical guidelines, mostly derived from experience (Carpenter et al. 1997), along with the design of experiments (DoE) approach (Politis et al. 2017; Grant et al. 2012).

Some practical guidelines can be useful for researchers who do not have extensive experience in formulation development, especially if they point out the main factors to consider when designing a freeze-dried formulation, that is, not only protein stability but also glass transition and collapse temperature, tonicity, route of administration, container, and so forth.

However, when the formulation is critical, it is important to use a more effective methodology. In this context, DoE and statistical analysis are commonly applied to formulation development, because they allow a quick and systematic evaluation of all the potential factors involved. With DoE, it is possible to evaluate the effect of each formulation factor on each monitored output, as well as potential interactions. Afterwards, the critical factors could be identified and adjusted, in order to optimize the formulation and guarantee its robustness.

However, a problem with current approaches for formulation design is that the molecular mechanisms at the basis of protein stabilization by excipients are generally not accessible by experimental techniques. This means that the formulation design is still mostly driven by experience and empirical observations, without real knowledge of the molecular-scale phenomena involved. Moreover, the number of factors to be considered is often huge, making the process extremely time-consuming.

A possible solution to this problem is to combine molecular-scale *in silico* modeling with experiments (Arsiccio et al. 2019; Pandya et al. 2018). More specifically, MD may be used to prescreen a wide number of potential excipients, identifying their stabilization mechanisms. This computational investigation would integrate the description offered by experimental data and would also make it possible to decrease the number of formulations to be experimentally studied, making the whole process more time efficient and cost-effective.

In the following, we will briefly introduce MD simulations and explain how they may help formulation development. Finally, some examples of applications will be provided, to show how MD results can drive the choice of suitable excipients for protein therapeutics.

2.4 MOLECULAR DYNAMICS AS A TOOL FOR FORMULATION DEVELOPMENT

Molecular simulations (Frenkel and Smit 2002; Allen and Tildesley 2017) are playing an increasing role in a wide number of fields, spanning physics, chemistry, biology, and life science. The key for their success is their ability of providing insight into problems that are often not directly accessible by experimental techniques, thus offering a better understanding of complex phenomena. In a common MD simulation, structure files describing the protein (in many cases a Protein Data Bank file) or the excipient molecule, and topology files listing the forces acting on each atom, are needed to set up a simulation. A simulation box is then built, where the protein is surrounded by the desired number of excipient molecules and solvated in water, and counter ions are also added to guarantee the neutrality of the system. After an energy minimization step, the system is equilibrated at the desired values of temperature and pressure. Finally, the production run can start, where Newton's law of motion is integrated over time for all the atoms in the simulation box, and the coordinates are written to an output file at regular intervals. A scheme of the process required for starting an MD simulation is shown in Figure 2.1.

The forces acting on each atom are defined by the force field, which describes both the nonbonded and bonded interactions in the system under investigation. The bonded interactions can be easily and quickly computed because they always involve the same atoms. For instance, a covalent bond between two atoms i and j, with positions defined by $\mathbf{r}_i$ and $\mathbf{r}_j$, is generally described according to the harmonic potential V_{ij},

$$V_{ij}\left(r_{ij}\right) = \frac{1}{2}k_{ij}\left(r_{ij} - b\right)^2, \quad r_{ij} = \left|\mathbf{r}_i - \mathbf{r}_j\right|, \tag{2.3}$$

where k_{ij} is the force constant and b the equilibrium distance between the two atoms.

The forces acting on angles θ_{ijk} and dihedrals Φ_{ijkl} formed within each molecule inside the simulation box are often defined according to the following potentials,

$$V_{ijk}\left(\theta_{ijk}\right) = \frac{1}{2}k_{ijk}\left(\theta_{ijk} - \theta_0\right)^2, \quad \theta_{ijk} = \cos^{-1}\left(\frac{\mathbf{r}_{ij}\cdot\mathbf{r}_{jk}}{r_{ij}r_{jk}}\right), \tag{2.4}$$

$$V_{ijkl}\left(\Phi_{ijkl}\right) = k_{ijkl}\left(1+\cos\left(\Phi_{ijkl} - \Phi_0\right)\right), \quad \Phi_{ijkl} = \cos^{-1}\left(\frac{(\mathbf{r}_{ij}x\mathbf{r}_{jk})\cdot(\mathbf{r}_{jk}x\mathbf{r}_{kl})}{\left|\mathbf{r}_{ij}x\mathbf{r}_{jk}\right|\left|\mathbf{r}_{jk}x\mathbf{r}_{kl}\right|}\right), \tag{2.5}$$

where k_{ijk} and k_{ijkl} are the force constants, while θ_0 and Φ_0 represent the equilibrium values. In some cases, the dihedral angles potential is expressed as a sum of powers of cosines, according to the Ryckaert-Belleman function,

$$V_{ijkl}\left(\Psi_{ijkl}\right) = \sum_{n=0}^{5} C_n\left(\cos\Psi_{ijkl}\right)^n, \quad \Psi_{ijkl} = \Phi_{ijkl} - 180°. \tag{2.6}$$

The nonbonded interactions are more difficult to compute, as the atoms involved change during the simulation time. This happens because the molecules are not fixed in space but are free to move within the simulation box. The nonbonded interactions include a repulsion term, a dispersion term, and a Coulomb term.

The repulsion and dispersion term are often combined in the Lennard-Jones (or 6–12) interaction,

$$V_{ij}\left(r_{ij}\right) = \frac{C_{12}}{r_{ij}^{12}} - \frac{C_6}{r_{ij}^6} = 4\varepsilon\left(\left(\frac{\sigma}{r_{ij}}\right)^{12} - \left(\frac{\sigma}{r_{ij}}\right)^6\right), \tag{2.7}$$

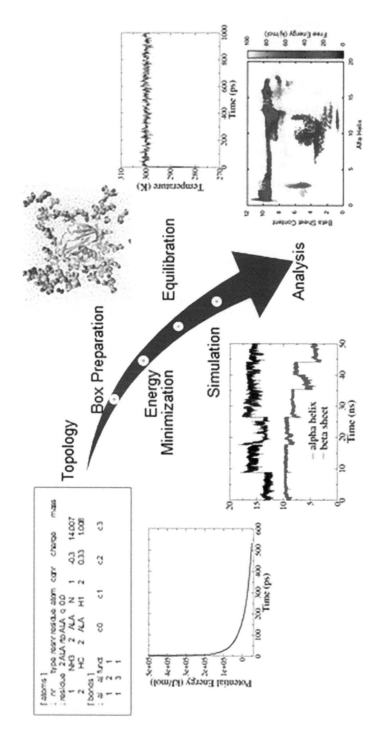

FIGURE 2.1   Scheme of the process required for starting a MD simulation.

where the parameters C_{12} and C_6 or σ and ε depend on pairs of atom types.

The Buckingham (or exp-6) potential may also be used,

$$V_{ij}\left(r_{ij}\right) = A_{ij}e^{-B_{ij}r_{ij}} - \frac{C_{ij}}{r_{ij}^6}. \tag{2.8}$$

In addition, atoms having charges q_i and q_j also act through the Coulomb term,

$$V_{ij}\left(r_{ij}\right) = \frac{1}{4\pi\epsilon_0}\frac{q_iq_j}{r_{ij}}, \tag{2.9}$$

where ε_0 is the absolute permittivity of free space.

Some of the most common all-atom, or united-atom, force fields are AMBER (Weiner and Kollman 1981), CHARMM (Mackerell et al. 1998), GROMOS (Scott et al. 1999), and OPLS (Jorgensen et al. 1996). All-atom force fields provide parameters for every single atom within the system, while united-atom ones provide parameters for all atoms except nonpolar hydrogens. The resulting description of the system is extremely detailed, resulting in the generation of a huge amount of information about the system but at a high computational cost. In fact, the main problem of atomistic MD is represented by the limited time and length scales that can be simulated.

A possible approach to reduce the computational cost is coarse-graining. In this case, the number of degrees of freedom in the model is reduced, allowing simulations of larger systems for longer times (Barnoud and Monticelli 2015). To achieve this, typically whole groups of atoms are represented as single beads, and the coarse-grained force field describes their interaction. However, approximations must be made, and the resulting description of the system under investigation is often less accurate.

In addition to this, the approximations that result from coarse-graining may be adequate only for a limited selection of molecules. Therefore, this type of approach may be difficult to implement for the simulation of typical formulations, where the chemical nature of the excipients used is often very broad.

Another option to overcome the timescale limitation of computer simulations consists in the use of enhanced sampling methods (Abrams and Bussi 2014). Some enhanced sampling approaches rely on the identification of a few order parameters, or collective variables (CVs), which are critical for the system of interest and whose fluctuations are therefore enhanced by a bias during the simulation time. A second class of techniques consists in simulating multiple replicas of the same system at different temperature, Hamiltonian or pressure, and then randomly exchanging the configurations with a given acceptance probability.

It is out of the scope of this chapter to provide detailed information about the different types of MD simulations. Therefore, we will only explain how some of the available techniques could be used to improve formulation design, mostly focusing on those approaches in which the authors have direct experience. The interested reader may find a more detailed description of the available MD simulation techniques in the studies by Abrams and Bussi (2014), Allen et al. (2009), Hollingsworth and Dror (2018), Morriss-Andrews and Shea (2015), Valsson et al. (2016), and references therein.

2.4.1 Use of Molecular Dynamics to Study Excipient-Protein Interactions

Molecular dynamics is a valuable tool for studying the interactions between a model protein and typical pharmaceutical excipients. For this type of simulation, the time scales that can be reached by classical MD simulations may be sufficient. Therefore, the use of enhanced sampling techniques is often not necessary.

A typical simulation box, in this case, would include a protein, solvated with water and surrounded by a given number of excipient molecules (see Figure 2.2).

The MD trajectory could be then analyzed to identify and quantify the mechanisms of protein stabilization by different osmolytes. For instance, molecular-level simulations allow evaluation of preferential interaction of proteins, in terms of either Kirkwood-Buff integrals or preferential exclusion coefficient Γ_{23}, with either water or cosolvents (Shukla et al. 2009; Ganguly and van der Vegt 2013). The number and strength of the hydrogen bonds formed within the simulation box, which are related to the formation of hydrogen-bonded clusters or to the degree of kinetic stabilization of the protein, can also be easily computed. In the past, this type of information has been computed for different classes of osmolytes, including arginine (Shukla and Trout 2010) or arginine-glutamate mixtures (Shukla and Trout 2011), trimethylamine-N-oxide (TMAO) and urea (Sarma and Paul 2013; Ganguly et al. 2018), trehalose (Cottone et al. 2002; Cordone et al. 2007; Lerbret et al. 2012; Corradini et al. 2013; Paul and Paul 2015) and other sugars (Lerbret et al. 2007;

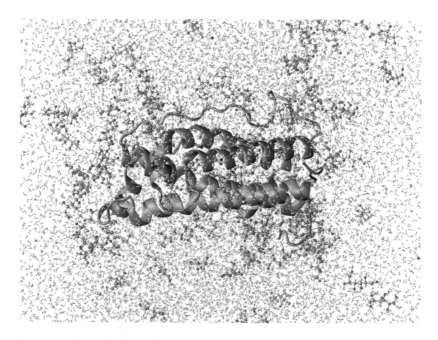

FIGURE 2.2 Example of an MD simulation box, where native human growth hormone is surrounded by sucrose molecules and solvated in water.

Cloutier et al. 2018). A wide range of osmolytes typically used in freeze-dried formulations has been recently investigated with this technique (Arsiccio and Pisano
2017; Arsiccio and Pisano 2018a-d; Arsiccio et al. 2018).

In some cases, however, the effect of an excipient on protein conformational
changes may be the main interest of the *in silico* modeling, and these conformational changes may occur on timescales that cannot be easily reached by classical
MD simulations. In these cases, the use of enhanced sampling techniques should be
considered, as they can help speed up the transitions between different protein structures. In particular, enhanced sampling techniques are extremely useful in the case
of energy landscapes that feature many minima and barriers between these minima
that can be difficult to cross.

Among the different possible approaches, the so-called parallel-tempering replica exchange molecular dynamics (REMD) is often used (Hansmann 1997; Sugita
and Okamoto 1999). REMD simulations enhance the sampling by running several
independent replicas of the system of interest at different temperatures, and periodi-
cally exchanging the coordinates between the replicas with the following acceptance
probability,

$$\alpha = \min\left(1, e^{\left(\frac{1}{k_B T_j} - \frac{1}{k_B T_i}\right)\left[U_j - U_i\right]}\right),$$ (2.10)

where we have imagined an exchange between two replicas at temperatures T_i and
T_j and having energies U_i and U_j. Typically, the set of replicas is constructed so that
the replica at the lowest temperature represents the ensemble from which sampling
is wanted, while the highest temperature is chosen so that the barriers in the energy
landscape will be crossed over accessible simulation time scales.

Another technique that is frequently used is the so-called metadynamics
(Valsson et al. 2016), which belongs to the class of enhanced sampling techniques
relying on the identification of collective variables. Different versions of metadynamics exist, but in all cases an external history-dependent bias potential V is constructed in the space of the selected collective variables s_i, which are functions of
the atomic coordinates $s_i(\mathbf{R})$, in order to push the system away from local minima
and promote the sampling of new regions of the phase space. The potential is built
as a sum of Gaussian kernels, having height w_i and width σ_i, deposited along the
trajectory,

$$V(s_i, t) = \int_0^t w_i(t') \exp\left(-\frac{\left(s_i(\mathbf{R}) - s_i(\mathbf{R}(t'))\right)^2}{2\sigma_i^2}\right) dt'.$$ (2.11)

In the following, an overview of the main results obtained from the application of
MD simulations to formulation design will be provided. We will neglect technical
details and focus our attention on those findings that may be most useful for formulation scientists in the field of freeze-drying.

2.5   MOLECULAR DYNAMICS FOR FORMULATION DEVELOPMENT: EXAMPLES OF APPLICATION

2.5.1   THERMODYNAMIC STABILIZATION: A NEW INTERPRETATION OF THE PREFERENTIAL EXCLUSION THEORY

The coordination number of water and excipient molecules around the protein surface can be easily extracted from a MD trajectory, and the preferential exclusion, or interaction, of the excipient with the protein can be then straightforwardly computed. This was done for a wide range of excipient molecules. For instance, the counteracting effect of TMAO against urea-induced protein denaturation was related to the inhibition of protein-urea preferential interaction in the presence of TMAO (Ganguly et al. 2018), and the extent of this counteraction was observed to depend heavily on the amino acid composition of the peptide. Preferential hydration of lysozyme was observed in glassy matrices formed by trehalose (Lerbret et al. 2012), and trehalose was found to be excluded from the surface of carboxy-myoglobin as well (Cottone et al. 2002). The positive effect of trehalose on protein stability was also related to its ability to significantly slow down the relaxation of protein-water hydrogen bonds (Lerbret et al. 2012). Exclusion of a large number of excipient molecules, including sugars, polyols, and amino acids, from the surface of human growth hormone (hGH) was observed (Arsiccio and Pisano 2017).

A different possible interpretation of the preferential exclusion theory was also recently proposed (Arsiccio et al. 2019). An ideal experiment was imagined, where a protective osmolyte was added to an aqueous solution of a protein. If the osmolyte is excluded from the protein surface, the protein stability should increase with excipient concentration because the unfolding process would be thermodynamically unfavorable (Figure 2.3, red curve). However, in the limit of infinite excipient concentration, protein stabilization by preferential exclusion would be impossible, as the amount of residual water would be negligible and the solution could be considered homogeneous. Therefore, the preferential exclusion theory would predict a decrease

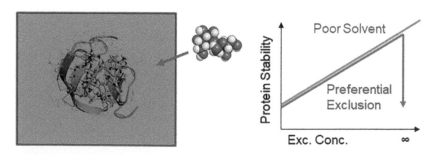

FIGURE 2.3   Schematic of an ideal experiment, where a protective osmolyte is added to a protein aqueous solution (left), and the resulting protein stability is monitored as function of the excipient concentration (right). The protein stability curves as predicted by the preferential exclusion theory (red line) or the poor solvent hypothesis (blue line) are shown.

in protein stability in the limit of infinite excipient concentration, which seems an unphysical and unrealistic result.

However, if we slightly change this perspective, this apparent paradox may be easily solved. Let us imagine that the protein is stabilized in a poor solvent, which favors the collapsed native conformation against the expanded unfolded ones. Let us further suppose that the osmolyte is a poorer solvent for the peptide chain than water. In this case (Figure 2.3, blue curve), protein stability would monotonically increase with excipient concentration, in the limit of infinite concentration as well, as we would continuously move from a good solvent, such as pure water, to a bad solvent, such as the osmolyte. In this framework, preferential exclusion would be a reflection of the unfavorable protein-osmolyte interactions, and this "poor solvent theory" would also have the advantage of unifying the theories for protein folding and for polymers stability in solutions.

However, proteins are a very heterogeneous class of polymers, as a huge number of possible sequences could result from the assembly of the 20 amino acids. Because of this, a good solvent for specific patches on the protein surface could be, at the same time, a poor solvent for other regions of the same protein. In line with these considerations, it was observed that some excipients were excluded only from specific side-chain sequences of hGH and LDH (Arsiccio and Pisano 2018b). Interestingly, it was also observed that the interaction of the osmolytes with these sequences was the key factor for protein stabilization. This suggests that specific regions on the protein surface, rather than the whole molecule, may drive the unfolding process during freeze-drying, as well as its inhibition by means of excipients.

2.5.2 Kinetic Stabilization: Role of Hydrogen Bonds

Kinetic stabilization is driven by the formation of a viscous, glassy matrix, where a compact hydrogen-bonding network prevents the protein motions responsible for unfolding. Molecular dynamics is a valuable tool that can help to provide insight into this mechanism of protein stabilization, as well.

MD simulations revealed that many common excipient molecules, such as the disaccharides sucrose and trehalose, could hydrogen bond to the polar residues exposed onto the protein surface. However, because of their generally larger molecular size with respect to water, a void space was formed between the osmolytes and the protein, where water molecules were entrapped. A new interpretation was therefore proposed for the most well-known mechanisms of stabilization in the dried state, i.e., vitrification, water replacement, and water entrapment (Arsiccio and Pisano 2018a). According to the MD simulations, these apparently different and conflicting mechanisms may actually be concurrent aspects of the same phenomenon, that is, the formation of a dense, compact hydrogen-bonding network. This network results in the formation of a cage structure of excipient molecules around the protein, and thus, to water entrapment, but the protective osmolytes can also hydrogen bond to some residues of the protein, generally the polar ones, in-line with the water replacement theory. Formation of this network also produces a sharp increase in viscosity, according to the vitrification scenario. A good excipient for stabilization in the dried state should therefore have a large number of hydrogen-bonding sites, and these should be easily accessible on the osmolyte surface (Arsiccio and Pisano 2017).

As previously discussed, the vitrification hypothesis generally associates a higher viscosity to slower chemical and physical degradation rates of proteins. Viscosity is closely related to the slowest and strongly temperature-dependent motions of a glass, known as α-relaxation processes (Yoshioka and Aso 2007). However, some studies seem to indicate that the protein stability in the dried state is not directly related to the α-relaxation time (Rossi et al. 1997; Wang et al. 2009), while the faster β-relaxation processes should be strongly related with the protein degradation rates (Wang et al. 2009; Cicerone and Douglas 2012; Cicerone et al. 2015).

With incoherent inelastic neutron scattering it is possible to measure the mean square displacement $<u^2>$ of hydrogen atoms in lyophilized products, which is a measure of the β-relaxation dynamics (Cicerone and Soles 2004). It was shown that $<u^2>$ is strongly related with protein stability in sugar glasses (Wang et al. 2009; Cicerone and Douglas 2012), and the value of $<u^2>$ was also computed from MD trajectories, for the case of different formulations including sugars (trehalose, cellobiose, sucrose, and mannitol) amino acids (histidine) or the β-cyclodextrin (Arsiccio et al. 2019). This analysis made it possible to conclude that the disaccharides should provide the best kinetic stabilization of the protein structure, while mannitol, histidine, and cyclodextrin should be much less effective.

In the context of kinetic stabilization, it is also important to take into account the possible interactions between the different components of the formulation. For instance, the addition of common buffers, such as phosphate and citrate buffers, was observed to affect the hydrogen-bonding network formed by excipients, such as sucrose and trehalose. As a consequence, the glass transition temperature of the formulations could significantly change (Weng and Elliott 2015), and there may be an effect on protein stability, as well.

In the case of hGH, it was observed that buffers had a negative effect on protein stabilization by excipients such as trehalose that form a highly structured hydrogen-bonding matrix. In contrast, the effect was beneficial for other osmolytes, such as sucrose, that mostly have a thermodynamic effect on protein stability (Arsiccio and Pisano 2018c).

Similar phenomena of interaction between formulation components were observed in the case of amino acids as excipients. MD simulations were used to investigate the role of arginine in the inhibition of protein aggregation (Shukla and Trout 2010), and it was proposed that the self-association of arginine molecules could be at the basis of the observed stabilization. The hydrogen bonds formed in the arginine clusters were stronger than those between arginine and water, and the formation of these large, stable clusters crowded out protein-protein interactions.

By contrast, when glutamic acid was also added to the protein-arginine formulation (Shukla and Trout 2011), the protein solubility was enhanced because of the increase in the number of hydrogen-bonded excipient molecules around the protein. The presence of these additional molecules around the protein resulted in enhanced crowding, suppressing protein aggregation.

From these examples it is therefore clear that the synergistic effects between excipients should also be carefully considered when designing a formulation.

2.5.3   KINETICS VS. THERMODYNAMICS: HOW TO DESIGN THE OPTIMAL FORMULATION

As previously mentioned, the mechanisms of protein stabilization change between the freezing and drying steps of freeze-drying, mostly because of water removal. During freezing, at least before the last stages of cryoconcentration, the water content is still high enough for the thermodynamic mechanisms to be dominant. The kinetic mechanisms prevail, by contrast, during the drying steps of lyophilization. This means that an optimal freeze-dried formulation should include both a good cryoprotectant, capable of providing thermodynamic stabilization, and an efficient lyoprotectant, which should be able to entrap the therapeutic protein in a compact matrix and kinetically hinder denaturing phenomena.

As explained in the previous sections, MD simulations could easily provide information about both thermodynamic and kinetic mechanisms of stabilization, making it possible to build a two-dimensional graphic representation of excipients efficiency (Figure 2.4). In Figure 2.4, excipients that are on the far right of the graph provide the best thermodynamic stabilization (cryoprotection), while those in the upper part should be the best lyoprotectants (Arsiccio and Pisano 2018e).

It is clear from this graph that not all the excipients are equally effective and that the disaccharides, in particular, should perform significantly better than polyols and amino acids. Among the disaccharides, however, a tiny difference can be observed, with sucrose and lactose excelling in thermodynamic stabilization and cellobiose and trehalose in kinetic protection.

This graph may prove extremely useful for people involved in formulation design, as it would allow selection of a formulation capable of providing the best stabilization during both the freezing and the drying steps of the lyophilization process.

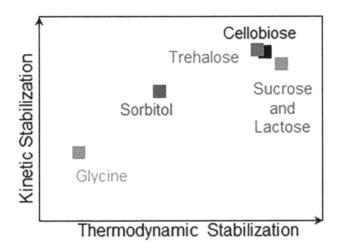

FIGURE 2.4   Graphic representation of the ability of common pharmaceutical excipients in providing either kinetic (*y*-axis) or thermodynamic (*x*-axis) stabilization. From the paper by Arsiccio and Pisano (2018e), with modifications.

According to the graph, for instance, a mixture of sucrose and trehalose combines an efficient cryoprotectant with a good lyoprotectant and should therefore provide a slightly better stabilization than a sucrose-only or a trehalose-only formulation. These considerations were also supported by experimental data in the case of lactate dehydrogenase as model protein (Arsiccio et al. 2019) and show how the identification of the molecular mechanisms responsible for protein stabilization could practically guide the choice of excipients.

2.5.4   PROTEIN STABILIZATION BY SURFACTANTS: EVIDENCE OF AN ORIENTATION-DEPENDENT MECHANISM

Finally, we would like to focus our attention on the role of surfactants, a class of excipients that is commonly used to prevent surface-induced denaturation. In particular, the ice-water, glass-water, and air-water interfaces are among the most commonly encountered in a freeze-drying process.

MD simulations (Arsiccio et al. 2018) revealed that surfactants stabilize proteins in presence of the denaturing air-water or ice-water interfaces. In the case of the air-water surface, the surfactant molecules preferentially located at the interface, inhibiting protein adsorption and stabilizing the native fold. The behavior was found to be different at the ice-water surface, where the surfactant molecules preferentially clustered around the protein, with their hydrophilic heads oriented toward the protein. This orientation had the effect of stabilizing the native structure by inhibiting the exposure of hydrophobic residues, which generally occurs upon unfolding.

In the presence of a hydrophilic silica-water surface, the surfactant molecules coated the interface, but also directly interacted with the protein. In this case, the tails-toward-the-protein configuration was favored, which stabilized partially unfolded states of the protein, characterized by a larger nonpolar surface area.

A similar orientation-dependent mechanism of the surfactants was also observed in the bulk solution (Arsiccio and Pisano 2018d). In this case, Tween 20 was found to prevent the self-association of hGH because of the formation of protein-surfactant complexes. Surfactant molecules were also observed to orient with their hydrophilic heads in the direction of unfolded hGH, fostering the refolding process.

2.6   CONCLUSIONS

We would like to conclude this chapter briefly summarizing what we think should be the main take-home message.

We have shown that molecular dynamics simulations can help to clarify the mechanisms at the basis of protein stabilization by excipients, with some examples of practical application. This information could then be used to guide the choice of an optimal freeze-dried formulation for therapeutic proteins.

The type of information that can be obtained from MD simulations is complementary, not alternative, to experimental tests. *In silico* modeling can help explain experimental results, and reduce the number of experimental tests to be performed, but it would hardly be able to reproduce the complexity of reality. At the same time, the molecular details provided by MD simulations are not accessible by current experimental techniques.

In this framework, it is clear that the optimal solution would be to combine *in silico* modeling and experimental investigation. Integration of the two approaches would provide a powerful tool, which could boost formulation design.

The hope of the authors is that the present chapter may make even a modest contribution to progress along these lines.

REFERENCES

Abrams, C., and G. Bussi. 2014. Enhanced sampling in molecular dynamics using metadynamics, replica-exchange, and temperature-acceleration. *Entropy* 16:163–199.

Allen, M. P., and D. J. Tildesley. 2017. *Computer simulation of liquids.* Oxford: Oxford University Press.

Allen, R. J., Valeriani, C., and P. R. ten Wolde. 2009. Forward flux sampling for rare event simulations. *J. Phys.: Condens. Matter* 21:463102.

Al-Hussein, A., and H. Gieseler. 2012. The effect of mannitol crystallization in mannitol – sucrose systems on LDH stability during freeze-drying. *J. Pharm. Sci.* 101:2534–2544.

Anchordoquy, T. J., and J. F. Carpenter. 1996. Polymers protect lactate dehydrogenase during freeze-drying by inhibiting dissociation in the frozen state. *Arch. Biochem. Biophys.* 332:231–238.

Arakawa, T., and S. N. Timasheff. 1982. Stabilization of protein structure by sugars. *Biochemistry* 21:6536–6544.

Arsiccio, A., and R. Pisano. 2017. Stability of proteins in carbohydrates and other additives during freezing: The human growth hormone as a case study. *J. Phys. Chem. B* 121:8652–8660.

Arsiccio, A., and R. Pisano. 2018a. Water entrapment and structure ordering as protection mechanisms for protein structural preservation. *J. Chem. Phys.* 148:055102.

Arsiccio, A., and R. Pisano. 2018b. Clarifying the role of cryo- and lyo-protectants in the biopreservation of proteins. *Phys. Chem. Chem. Phys.* 20:8267–8277.

Arsiccio, A., and R. Pisano. 2018c. The preservation of lyophilized human growth hormone activity: How do buffers and sugars interact? *Pharm. Res.* 35:131.

Arsiccio, A., and R. Pisano. 2018d. Surfactants as stabilizers for biopharmaceuticals: An insight into the molecular mechanisms for inhibition of protein aggregation. *Eur. J. Pharm. Biopharm.* 128:98–106.

Arsiccio, A., and R. Pisano. 2018e. Design of the formulation for therapeutic proteins – How to improve stability of drugs during freezing and in the dried state. *Chem. Today* 36:16–18.

Arsiccio, A., McCarty, J., Pisano, R., and J.-E. Shea. 2018. Effect of surfactants on surface-induced denaturation of proteins: Evidence of an orientation-dependent mechanism. *J. Phys. Chem. B* 122:11390–11399.

Arsiccio, A., Paladini, A., Pattarino, F., and R. Pisano. 2019. Designing the optimal formulation for biopharmaceuticals: A new approach combining molecular dynamics and experiments. *J. Pharm. Sci.* 108:431–438.

Bam, N. B., Randolph, T. W., and J. L. Cleland. 1995. Stability of protein formulations: Investigation of surfactant effects by a novel EPR spectroscopic technique. *Pharm. Res.* 12:2–11.

Bam, N. B., Cleland, J. L., and T. W. Randolph. 1996. Molten globule intermediate of recombinant human growth hormone: Stabilization with surfactants. *Biotechnol. Prog.* 12:801–809.

Bam, N. B., Cleland, J. L., Yang, J., et al. 1998. Tween protects recombinant human growth hormone against agitation-induced damage via hydrophobic interactions. *J. Pharm. Sci.* 87:1554–1559.

Barnoud, J., and L. Monticelli. 2015. Coarse-grained force fields for molecular simulations. *Methods. Mol. Biol.* 1215:125–149.

Belton, P. S., and A. M. Gil. 1994. IR and Raman spectroscopic studies of the interaction of trehalose with hen egg white lysozyme. *Biopolymers* 34:957–961.

Ben-Naim, A. 2006. *Molecular theory of solutions.* New York: Oxford University Press.

Carpenter, J. F., and J. H. Crowe. 1989. An infrared spectroscopic study of the interactions of carbohydrates with dried proteins. *Biochemistry* 28:3916–3922.

Carpenter, J. F., Crowe, J. H., and T. Arakawa. 1990. Comparison of solute-induced protein stabilization in aqueous solution and in frozen and dried state. *J. Dairy Sci.* 73:3627–3636.

Carpenter, J. F., Prestrelski, S. J., and T. Arakawa. 1993. Separation of freezing and drying-induced denaturation of lyophilized proteins using stress-specific stabilization. I. Enzymatic activity and calorimetric studies. *Arch. Biochem. Biophys.* 303:456–464.

Carpenter, J. F., Pikal, M. J., Chang, B. S., and T. W. Randolph. 1997. Rational design of stable lyophilized protein formulations: Some practical advice. *Pharm. Res.* 14:969–975.

Chang, B. S., Kendrick, B. S., and J. F. Carpenter. 1996. Surface-induced denaturation of proteins during freezing and its inhibition by surfactants. *J. Pharm. Sci.* 85:1325–1330.

Cicerone, M. T., and C. L. Soles. 2004. Fast dynamics and stabilization of proteins: Binary glasses of trehalose and glycerol. *Biophys. J.* 86:3836–3845.

Cicerone, M. T., and J. F. Douglas. 2012. β-relaxation governs protein stability in sugar-glass matrices. *Soft Matter* 8:2983–2991.

Cicerone, M. T., Pikal, M. J., and K. K. Qian. 2015. Stabilization of proteins in solid form. *Adv. Drug. Deliv. Rev.* 93:14–24.

Cloutier, T., Sudrik, C., Sathish, H. A., and B. L. Trout. 2018. Kirkwood-Buff-derived alcohol parameters for aqueous carbohydrates and their application to preferential interaction coefficient calculations of proteins. *J. Phys. Chem. B* 122:9350–9360.

Cordone, L., Cottone, G., and S. Giuffrida. 2007. Role of residual water hydrogen bonding in sugar/water/biomolecule systems: A possible explanation for trehalose peculiarity. *J. Phys.: Condens. Matter* 19:205110.

Corradini, D., Strekalova, E. G., Stanley, H. E., and P. Gallo. 2013. Microscopic mechanism of protein cryopreservation in an aqueous solution with trehalose. *Sci. Rep.* 3:1218.

Cottone, G., Ciccotti, G., and L. Cordone. 2002. Protein – trehalose – water structures in trehalose coated carboxy-myoglobin. *J. Chem. Phys.* 117:9862–9666.

Crowe, J. H., Crowe, L. M., and D. Chapman. 1984. Preservation of membranes in anhydrobiotic organisms: The role of trehalose. *Science* 223:701–703.

Deechongkit, S., Wen, J., Narhi, L. O., et al. 2009. Physical and biophysical effects of polysorbate 20 and 80 on darbepoetin alfa. *J. Pharm. Sci.* 98:3200–3217.

Franks, F. 1994. Long-term stabilization of biologicals. *Biotechnology* 12:253–256.

Franks, F. 1995. Protein destabilization at low temperatures. *Adv. Protein Chem.* 46:105–139.

Frenkel, D., and B. Smit. 2002. *Understanding molecular simulation.* New York: Academic.

Ganguly, P., and N. F. A. van der Vegt. 2013. Convergence of sampling kirkwood–buff integrals of aqueous solutions with molecular dynamics simulations. *J. Chem. Theory Comput.* 9:1347–1355.

Ganguly, P., Boserman, P., van der Vegt, N. F. A., and J.-E. Shea. 2018. Trimethylamine N-oxide counteracts urea denaturation by inhibiting protein – urea preferential interaction. *J. Am. Chem. Soc.* 140:483–492.

Gómez, G., Pikal, M. J., and N. Rodríguez-Hornedo. 2001. Effect of initial buffer composition on pH changes during far-from-equilibrium freezing of sodium phosphate buffer solutions. *Pharm. Res.* 18:90–97.

Grant, Y., Matejtschuk, P., Bird, C., Wadhwa, M., and P. A. Dalby. 2012. Freeze drying formulation using microscale and design of experiment approaches: A case study using granulocyte colony-stimulating factor. *Biotechnol. Lett.* 34:641–648.

Graziano, G., Catanzano, F., Riccio, A., and G. Barone. 1997. A reassessment of the molecular origin of cold denaturation. *J. Biochem.* 122:395–401.

Green, J. L., and C. A. Angell. 1989. Phase relations and vitrification in saccharide-water solutions and the trehalose anomaly. *J. Phys. Chem.* 93:2880–2882.

Hagen, S. J., Hofrichter, J., and W. A. Eaton. 1995. Protein reaction kinetics in a room-temperature glass. *Science* 269:959–962.

Hansmann, U. H. 1997. Parallel tempering algorithm for conformational studies of biological molecules. *Chem. Phys. Lett.* 281:140–150.

Heller, M. C., Carpenter, J. F., and T. W. Randolph. 1997. Manipulation of lyophilization induced phase separation: Implications for pharmaceutical proteins. *Biotechnol. Prog.* 13:590–596.

Hollingsworth, S. A., and R. O. Dror. 2018. Molecular dynamics simulations for all. *Neuron* 99:1129–1143.

Jiskoot, W., Randolph, T. W., Volkin, D. B., et al. 2012. Protein instability and immunogenicity: Roadblocks to clinical application of injectable protein delivery systems for sustained release. *J. Pharm. Sci.* 101:946–954.

Jones, L. S., Randolph, T. W., Kohnert, U., et al. 2001. The effects of Tween 20 and sucrose on the stability of anti-L-selectin during lyophilization and reconstitution. *J. Pharm. Sci.* 90:1466–1477.

Jorgensen, W. L., Maxwell, D. S., and J. Tirado-Rives. 1996. Development and testing of the OPLS all-atom force field on conformational energetics and properties of organic liquids. *J. Am. Chem. Soc.* 118:11225–11236.

Kirkwood, J. G., and F. P. Buff. 1951. The statistical mechanical theory of solutions. I. *J. Chem. Phys.* 19:774.

Lee, H. J., McAuley, A., Schilke, K. F., and J. McGuire. 2011. Molecular origins of surfactant-mediated stabilization of protein drugs. *Adv. Drug. Deliv. Rev.* 63:1160–1171.

Lerbret, A., Affouard, F., Hedoux, A., et al. 2012. How strongly does trehalose interact with lysozyme in the solid state? Insights from molecular dynamics simulation and inelastic neutron scattering. *J. Phys. Chem. B* 116:11103–11116.

Lerbret, A., Bordat, P., Affouard, F., Hedoux, A., Guinet, Y., and M. Descamps. 2007. How do trehalose, maltose, and sucrose influence some structural and dynamical properties of lysozyme? Insight from molecular dynamics simulations. *J. Phys. Chem. B* 111:9410–9420.

Liu, W., Wang, D. Q., and S. L. Nail. 2005. Freeze-drying of proteins from a sucrose-glycine excipient system: Effect of formulation composition on the initial recovery of protein activity. *AAPS PharmSciTech.* 6:150–157.

Lopez, C. F., Darst, R. K., and P. J. Rossky. 2008. Mechanistic elements of protein cold denaturation. *J. Phys. Chem. B* 112:5961–5967.

MacKerell, A. D., Bashford, D., Bellott, M., et al. 1998. All-atom empirical potential for molecular modeling and dynamics studies of proteins. *J. Phys. Chem. B* 102:3586–3616.

Matysiak, S., Debenedetti, P. G., and P. J. Rossky. 2012. Role of hydrophobic hydration in protein stability: A 3D water-explicit protein model exhibiting cold and heat denaturation. *J. Phys. Chem. B* 116:8095–8104.

Mi, Y., Wood, G., and L. Thoma. 2004. Cryoprotection mechanisms of polyethylene glycols on lactate dehydrogenase during freeze-thawing. *AAPS J.* 6:e22.

Morriss-Andrews, A., and J.-E. Shea. 2015. Computational studies of protein aggregation: Methods and applications. *Ann. Rev. Phys. Chem.* 66:643–666.

Murase, N., and F. Franks. 1989. Salt precipitation during the freeze-concentration of phosphate buffer solutions. *Biophys. Chem.* 34:293–300.

Ohtake, S., Kita, Y., and T. Arakawa. 2011. Interactions of formulation excipients with proteins in solution and in the dried state. *Adv. Drug Deliv. Rev.* 63:1053–1073.

Pandya, A., Howard, M. J., Zloh, M., and P. A. Dalby. 2018. An evaluation of the potential of NMR spectroscopy and computational modelling methods to inform biopharmaceutical formulations. *Pharmaceutics* 10:165.

Paul, S., and S. Paul. 2015. Molecular insights into the role of aqueous trehalose solution on temperature-induced protein denaturation. *J. Phys. Chem. B* 119:1598–1610.

Pikal, M. J. 2010. Mechanisms of protein stabilization during freeze-drying storage: The relative importance of thermodynamic stabilization and glassy state relaxation dynamics. In *Freeze-Drying/Lyophilization of pharmaceutical and biological products,* 3rd Edition, ed. L. Rey and J. C. May, 198–232. London: Informa Healthcare.

Politis, S., Colombo, P., Colombo, G., and D. M. Rekkas. 2017. Design of experiments (DoE) in pharmaceutical development. *Drug. Dev. Ind. Pharm.* 43:889–901.

Privalov, P. L. 1990. Cold denaturation of proteins. *Crit. Rev. Biochem. Mol. Biol.* 25:281–305.

Randolph, T. W and L. S. Jones. 2002. Surfactant-protein interactions. In *Rational design of stable protein formulations,* ed. J. F. Carpenter, and M. C. Manning, 159–175. Boston, MA: Springer.

Roberts, C. J. 2014. Therapeutic protein aggregation: Mechanisms, design, and control. *Trends Biotechnol.* 32:372–380.

Rossi, S., Buera, M. P., Moreno, S., and J. Chirife. 1997. Stabilization of the restriction enzyme EcoRI dried with trehalose and other selected glass-forming solutes. *Biotechnol Prog.* 13:609–616.

Rupley, J. A., and G. Careri. 1991. Protein hydration and function. *Adv. Protein Chem.* 41:37–172.

Sarma, R., and S. Paul. 2013. Exploring the molecular mechanism of trimethylamine-N-oxide's ability to counteract the protein denaturing effects of urea. *J. Phys. Chem. B* 117:5691–5704.

Scott, W. R. P., Hünenberger, P. H., Tironi, I. G., et al. 1999. The GROMOS biomolecular simulation program package. *J. Phys. Chem. A* 103:3596–3607.

Shukla, D., and B. Trout. 2010. Interaction of arginine with proteins and the mechanism by which it inhibits aggregation. *J. Phys. Chem. B* 114:13426–13438.

Shukla, D., and B. Trout. 2011. Understanding the synergistic effect of arginine and glutamic acid mixtures on protein solubility. *J. Phys. Chem. B* 115:11831–11839.

Shukla, D. Shinde, C., and B. L. Trout. 2009. Molecular computations of preferential interaction coefficients of proteins. *J. Phys. Chem. B* 113:12546–12554.

Strambini, G. B., and E. Gabellieri. 1996. Proteins in frozen solutions: Evidence of ice-induced partial unfolding. *Biophys. J.* 70:971–976.

Sugita, Y., and Y. Okamoto. 1999. Replica-exchange molecular dynamics method for protein folding. *Chem. Phys. Lett.* 314:141–151.

Timasheff, S. N. 1993. The control of protein stability and association by weak interactions with water: How do solvents affect these processes? *Ann. Rev. Biophys. Biomol. Struct.* 22:67–97.

Valsson, O., Tiwary, P., and M. Parrinello. 2016. Enhancing important fluctuations: Rare events and metadynamics from a conceptual viewpoint. *Ann. Rev. Phys. Chem.* 67:159–184.

van den Berg, L., and D. Rose. 1959. The effect of freezing on the pH and composition of sodium and potassium solutions: The reciprocal system KH_2PO_4-Na_2HPO_4-H_2O. *Arch. Biochem. Biophys.* 81:319–329.

Wang, W. 2000. Lyophilization and development of solid protein pharmaceuticals. *Int. J. Pharm.* 203:1–60.

Wang, W., and C. J. Roberts. 2010. Aggregation of therapeutic proteins. Hoboken, NJ: John Wiley & Sons.

Wang, B., Tchessalov, S., Cicerone, M. T., Warne, N. W., and M. J. Pikal. 2009. Impact of sucrose level on storage stability of proteins in freeze-dried solids: II. Correlation of aggregation rate with protein structure and molecular mobility. *J. Pharm. Sci.* 98:3145–3166.

Weiner, P. K., and P. A. Kollman. 1981. AMBER: Assisted model building with energy refinement. A general program for modeling molecules and their interactions. *J. Comput. Chem.* 2:287–303.

Weng, L., and G. D. Elliott. 2015. Distinctly different glass transition behaviors of trehalose mixed with Na_2HPO_4 or NaH_2PO_4: Evidence for its molecular origin. *Pharm. Res.* 32:2217–2228.

Yoshioka, S., and Y. Aso. 2007. Correlations between molecular mobility and chemical stability during storage of amorphous pharmaceuticals. *J. Pharm. Sci.* 96:960–981.

3 Established and Novel Excipients for Freeze-Drying of Proteins

Ivonne Seifert and Wolfgang Friess

CONTENTS

3.1 INTRODUCTION

About 90 lyophilised protein drug products are currently marketed in the United States or Europe. These biopharmaceuticals cover monoclonal antibodies, hormones, enzymes, vaccines, and fusion proteins (Walsh 2014; Gervasi et al. 2018). Proteins are commonly freeze-dried to overcome or reduce instabilities that may occur in liquid state. Stabilisers and excipients are added to form an amorphous matrix, in which the protein molecules are embedded and show enhanced stability (Wang et al. 2010a). Sucrose is the classically used amorphous matrix former. Two theories can explain its stabilising mechanism: the water replacement theory and the vitrification concept. The water replacement theory relies upon hydrogen bonding between the sugar and the protein molecules, thus replacing the water molecules at the protein surface. Thereby, the sugar molecules stabilise the native conformation of the protein (Prestrelski et al. 1993). According to the vitrification concept, protein mobility is substantially reduced in the glassy matrix, and reaction rates between protein molecules

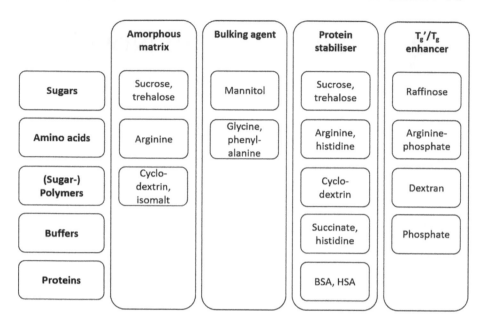

FIGURE 3.1 Exemplary excipients arranged in a matrix system according their chemical nature and functionality.

and between a protein molecule and, for example, water, oxygen, or other reactive species are decreased (Cicerone et al. 2015; Yoshioka and Aso 2007). More recent interpretations of these mechanisms, involving molecular dynamics simulations, are given and discussed in Chapter 2. In order to obtain a pharmaceutically acceptable parenteral product, excipients are required to provide protein stability but also isotonicity, physiologically compatible pH, and elegant appearance. In the following we discuss the typically used excipients as well as the potential of new excipients in freeze-drying of proteins. Some excipients fulfil multiple functions; for example, histidine acts as protein stabiliser, reducing aggregation and at the same time acting as buffer, and surfactants can stabilise the protein against freezing stress but also may accelerate reconstitution. In the following, the excipients are structured in a matrix by chemical nature and functionality to improve understanding, which is shown in Figure 3.1.

3.2 SUGARS AND POLYOLS

3.2.1 THE COMMONLY USED: SUCROSE, TREHALOSE, AND MANNITOL

The two disaccharides sucrose and trehalose are most commonly used in lyophilisation as protein stabilisers. They are approved for various parenteral routes. The small and flexible sugar molecules are able to cover the protein surface, forming hydrogen bonds with the protein molecules. Both form amorphous matrices, which are key for protein stabilisation. Thus, they act as both cryo- and lyoprotectant.

Sucrose is the most frequently used sugar in the freeze-drying of biopharmaceuticals. Besides the cryo- and lyoprotective effects of sucrose, a key feature is that it is a nonreducing disaccharide. A T_g' of $-32°C$ and a T_g of $74°C$ make sucrose favourable for both the lyophilisation process and storage stability (Heljo 2013). Solid contents of up to 10% (w/w) are suitable to form elegant cakes and to achieve isotonic products after reconstitution (Franks 1998). Nevertheless, the glycosidic bond of sucrose may be cleaved, forming the reducing monosaccharides glucose and fructose, which results in Maillard reactions with protein or amino group–carrying excipients (O'Brien 1996). Thus, even more stable stabilisers are pursued. Furthermore, there is interest in sugar matrices with a higher T_g' or T_c to achieve faster primary drying processes (Chang and Patro 2004). Recently, the pharmaceutical grade of sucrose was found to contain some nanoparticulate impurities (Weinbuch et al. 2015).

Trehalose has emerged as an alternative to sucrose in lyophilisation after the approval of the first lyophilisate stabilised with the trehalose Herceptin®, which contains trastuzumab. The glycosidic bond of this nonreducing disaccharide is more stable than that of sucrose (O'Brien 1996). Additionally, trehalose shows higher T_g' and T_g values than sucrose, with $-28°C$ and $119°C$, respectively (Heljo 2013). Trehalose can crystallise as dihydrate during freezing, which can negatively impact protein stability. Crystallization can be inhibited by combination with sucrose, for example (Sundaramurthi and Suryanarayanan 2010a). Furthermore, trehalose dihydrate formation occurs mainly during annealing, but upon drying, the dihydrate converts into the amorphous anhydrate (Sundaramurthi and Suryanarayanan 2010b).

The sugar alcohol mannitol is used on regular basis as bulking excipient for protein lyophilisates. Amorphous mannitol can help to stabilise proteins, such as LDH, β-galactosidase, and L-asparaginase, but has low T_g' values of $-32°C$ and $-25°C$ and a very low T_g of $12.6°C$ with a high tendency to crystallise (Cavatur et al. 2002; Yu et al. 1998). Upon crystallization mannitol loses its ability to stabilise proteins. Mannitol forms different polymorphs as well as a hemihydrate depending on the crystallization temperature, rate, and the annealing step during the lyophilisation process (Hawe and Friess 2006). The crystallization leads to a scaffold, which not only provides pharmaceutically elegant cake appearance but also enables process temperatures during primary drying above the T_g' or the collapse temperature T_c of the amorphous matrix. Exemplarily, at a mannitol to sucrose ratio of 4:1, primary drying can be run at $-10°C$ product temperature (Horn et al. 2018a). The amorphous sucrose matrix at this ratio stabilises proteins sufficiently (Passot et al. 2005).

3.2.2 OTHER SMALL SACCHARIDES AND POLYOLS

Sucrose is chosen widely for freeze-drying formulations because of its protein stabilising efficiency. However, amorphous sucrose-based formulations show often relatively low T_g values and can show stability problems when stored at higher temperatures or elevated humidities (te Booy et al. 1992). Various saccharides are under investigation because of their different physical properties and protein-stabilising effects (Pikal 1994).

Maltose showed the ability to maintain BSA and ovalbumin conformation during freeze-drying, whereas maltooligosaccharides had decreased stabilising effects with increasing saccharide units (Izutsu et al. 2004).

Screening different disaccharides regarding their stabilising effect of freeze dried β-galactosidase resulted in about 60% remaining activity after 90 days at 45°C for the following sugars: sucrose, trehalose, cellobiose (glucose and glucose), iso-maltulose (glucose and fructose), and melibiose (galactose and glucose). However, cellobiose, isomaltulose, and melibiose all belong to the group of reducing sugars and may therefore induce the Maillard reaction (Heljo 2013).

Owing to their higher molecular weight, trisaccharides require more solid content to reach isotonicity in freeze-drying formulations. This, on the one hand, may provide more amorphous matrix mass in which protein molecules are embedded and could be considered beneficial especially at high protein concentration to reduce interactions. On the other hand, a higher total solid content may slow down drying and delay reconstitution. Furthermore, large saccharides can provide high T_g values (Izutsu et al. 2004) but are reported to be less effective in protein stabilisation due to steric hinderance with increasing size (Tanaka et al. 1991). Melezitose, a nonreducing trisaccharide based on two glucose and one fructose unit and produced by plant eating insects, which shows a high T_g of 160°C, was evaluated for lyophilisation of blood and plasma proteins. The stabilising effect for factor VIII was better than trehalose during processing and after storage at 40°C for 4 weeks. Furthermore, melezitose provided good stability of factor IX, rFVIII, pFIX, and rHES-G-CSF lyophilisates over 6 months (Ivarsson et al. 2013, 2018; Mollmann et al. 2008).

Raffinose is composed of galactose, glucose, and fructose, but it is a nonreducing sugar owing to its chemical stability. It forms amorphous lyophilisates with a T_g' of −26°C and a T_g of 109°C. Annealing at −10°C results in the crystalline raffinose pentahydrate form, which was dehydrated during primary drying and eventually became amorphous. LDH activity was reduced in the annealed samples, although the final product containing 5% to 14% raffinose was amorphous (Chatterjee et al. 2005b; Heljo 2013). The stabilising potential of raffinose via water replacement appears to be inferior to sucrose. Whereas 100% raffinose resulted in a markedly higher T_g of 37°C at approximately 5% residual moisture than sucrose and raffinose/sucrose mixtures, the remaining LDH activity was higher with higher sucrose content upon storage at 44°C for 45 days (Davidson and Sun 2001).

While mannitol is used as a crystalline bulking agent, other polyols, which form an amorphous phase could act as cryo- and lyoprotectors. Furthermore, small polyols like glycerol and sorbitol can act as plasticisers, excellent hydrogen-bond formers, and void fillers in glasses (Cicerone and Soles 2004; Chang et al. 2005; Izutsu et al. 1994; Carpenter et al. 2002).

The plasticising effect of glycerol leads to a T_g decrease and a molecular mobility increase of a sugar matrix (Forney-Stevens et al. 2016; Cicerone and Soles 2004). Adding a small amount of glycerol to a trehalose formulation decreased the protective effect of trehalose to lysozyme during primary drying owing to the plasticiser behaviour. In secondary drying, glycerol showed an antiplasticising effect by forming strong hydrogen bonds with trehalose, thereby suppressing fast local dynamics of trehalose (Starciuc et al. 2017). Interestingly, adding small amounts of the plasticiser sorbitol to a sucrose formulation for an IgG1 resulted in a decrease in subvisible particle levels after lyophilisation as well as after subsequent storage at 40°C for 4 weeks, even at a relatively high sucrose to protein ratio of 10:1 (Davis et al. 2013).

The sugar alcohols maltitol, lactitol, and maltotriitol were able to form amorphous glasses during freeze-drying and prevented activity loss of LDH upon storage even at 50°C. They also protected bovine serum albumin from lyophilisation-induced secondary structure perturbation. In contrast, xylitol, a pentitol, and the hexitols sorbitol and mannitol collapsed or resulted in crystalline solids (Kadoya et al. 2010).

The sweetening agent and tabletting excipient isomalt consists of two stereoisomers, 6-O-α-D-glucopyranosyl-D-sorbitol and 1-O-α-D-glucopranosyl-D-mannitoldihydrate. Isomalt forms amorphous lyophilisates, which did not crystallise even at up to 16% residual moisture. Stereoisomer mixtures at four different ratios could not prevent a loss of LDH activity during freeze-drying, but preserved the enzymatic activity better than sucrose during long term stability testing (Sentko and Willibald-Ettle 2012; Tuderman et al. 2018). Overall, the polyols could be options for protein stabilisation, potentially in mixture with sucrose.

Meso-erythritol was identified as potential bulking agent for lyophilisation owing to its high crystallization propensity (Fujii et al. 2015), but it has not been tested with biopharmaceuticals.

3.2.3 OLIGO- AND POLYSACCHARIDES

Dextrans are oligo-glucosides, which stay amorphous during lyophilisation (Hancock and Zografi 1997; Allison et al. 2000). Pure 70 kDa and 6 kDa dextran show in solution T_g' values of −11°C and −14°C and T_g values of 167°C and 144°C, respectively. Adding 70 kDa dextran to trehalose at a 1:1 ratio led to an increase of T_g' from −28°C to −20°C and of T_g from 88°C to 110°C. However, pure 70 kDa dextran and dextran/trehalose combinations did not improve the stability of insulin, LDH, β-galactosidase, and HBsAg upon storage at 60°C (<10% r.h.) for 4 weeks as compared to trehalose (Tonnis et al. 2015). Along with other polymers, dextrans can cause liquid-liquid phase separation during freeze concentration when used as cryoprotectant (Randolph 1997; Izutsu et al. 1996).

The 4 kDa fructane inulin also results in higher T_g' and T_g values of −17°C and 154°C, as compared to −28°C and 121°C of trehalose, respectively. However, inulin did not improve storage stability of proteins at 60°C for 4 weeks as compared to trehalose, although it performed better than 70 kDA dextran (Tonnis et al. 2015).

Another polyglucoside, which can be used as an amorphous T_g' and T_g modifier, is maltodextrin, which is obtained by partial hydrolysis of starch. Five-percent maltodextrin with a dextrose equivalent (DE) of 5 or 8 showed a T_g' of −10°C and enables faster drying cycles (Passot et al. 2005). Formulations containing maltodextrin DE 5 or 8 and Polysorbate 80 or polyethyleneglycol PEG ensured stability of two *Clostridium difficile* toxins at ambient temperature (Passot et al. 2005). In general, the dextrans, inulin, and maltodextrin have chemically reactive end groups and should therefore be used with care in a protein formulation.

Whereas hydroxyethyl cellulose (HEC) 90 kDa, known as a gelling and thickening agent, provided better protection of LDH against freeze-thaw stress than sucrose, less activity was found after freeze-drying than with sucrose. The HEC lyophilisates had a high T_g of 160°C, but reconstitution took 40 min for 1% HEC (Al-Hussein and Gieseler 2015).

Hydroxyethyl starch (HES) shows similar benefits as dextran on the physical properties of lyophilised proteins (see 3.2.3). Whereas interleukin-11 (rhIL-11) lyophilisates with 2.5% 200 kDa HES as sole excipient showed high T_g, indicating potentially high storage stability, HES failed to inhibit lyophilisation and storage-induced unfolding of rhIL-11. The combination of 2.5% HES with 2.5% or 5% sucrose or trehalose formulation of rhIL-11 resulted in good storage stability through water replacement by the sugar molecules and the increased T_g values arising from HES (Garzon-Rodriguez et al. 2004). The stability of HESylated interferon α-2b was superior to PEGylated, indicating a beneficial effect of HES also in conjugated form (Liebner et al. 2015).

Cyclodextrins (CDs) are cyclic, nonreducing oligoglucosides and can be differentiated into three main classes, α-, β-, and γ-CDs, which are composed of six, seven, and eight glucose molecules, respectively. They are hydrophilic on the outside, have a rather hydrophobic core, and can be used to solubilise and stabilise poorly water-soluble compounds. Owing to systemic toxicity concerns and to increase water solubility, CDs are functionalised (Frijlink et al. 1991; Frank et al. 1976) and hydroxypropyl- and sulfobutyl-CDs are approved for parenteral products (Shiotani et al. 1995; Gould and Scott 2005).

With their hydrophobic pocket, CDs can interact with lipophilic amino acid side chains of proteins, ultimately reducing hydrophobic interactions. Thus CD derivates can reduce protein aggregation in liquid and could stabilise proteins during cooling or freezing of solutions and upon rehydration of lyophilisates (Serno et al. 2011).

CD derivates can increase T_g values; for example, freeze-dried β-CD with interleukin-2 (IL-2) had an increased T_g up to 108°C with 2.5%–3.5% RM, which might enhance protein stability during storage (Prestrelski et al. 1995; Santagapita et al. 2008). Furthermore, HP-β-CD provides a high collapse temperature T_c of −9°C (Bosch 2014; Meister et al. 2009).

At higher concentrations, HP-β-CD stabilises via protein-water replacement and vitrification (Iwai et al. 2007). Addition of HP-β-CD to trehalose reduced the aggregation of an IgG with an optimal mass ratio of trehalose to HP-β-CD of approximately 3.3 to 1 (Faghihi et al. 2017). Both pure trehalose and pure HP-β-CD kept the IgG aggregation level low in formulations containing 80% carbohydrate and 20% IgG (Faghihi et al. 2016). HP-α-CD, HP-β-CD, PM-β-CD, and HP-γ-CD preserved LDH activity during freeze-drying, with more than 80% preserved, compared to trehalose with 70% (Iwai et al. 2007).

Low HP-β-CD concentrations of about 0.1% provide a surfactant-like prevention of protein aggregation at interfaces. HP-β-CD may thus be an alternative to polysorbate or poloxamer in protein drug products, especially in case of a low specific surface area of the product (Geidobler 2014).

3.3   AMINO ACIDS

Amino acids are commonly used in protein liquid formulations as buffer systems, to stabilise the native protein structure, to enhance protein solubility, or to reduce viscosity at high concentration (Golovanov et al. 2004; Wang et al. 2015; Arakawa et al. 2007).

With respect to protein lyophilisates it is suggested that charged amino acids may be good void fillers of amorphous sucrose matrices. In this respect positively charged amino acids had an advantageous stabilising effect on human serum albumin (rHSA), potentially related to their larger molar volume, as compared to negatively charged amino acids (Forney-Stevens et al. 2016). Amino acids are known for their low T_g' values, for example $-62°C$ for glycine (Liu 2006) or $-42°C$ for L-arginine and $-47°C$ for L-arginine·HCl (Izutsu et al. 2016), which can be critical for the lyophilisation process (Remmele et al. 2012; Lueckel et al. 1998). Nevertheless, freeze-dried amino acids resulted often in high T_g values, for example about $90°C$ for L-arginine with citrate or about $100°C$ for L-histidine citrate, both around pH 6 (Izutsu et al. 2009a). Furthermore, amino acids like glycine and phenylalanine crystallise during freeze-drying and can therefore act as bulking agents (Horn et al. 2018b; Mattern et al. 1999; Kasraian et al. 1998; Lueckel et al. 1998). A glycine to raffinose or trehalose ratio of at least approximately 1.2 to 1.5 is required to render crystalline glycine, which enables manufacturing of elegant lyophilisates even at a primary drying temperature $10°C$ above T_g'. The combination of sugar and bulking agent prevented loss of LDH activity during freeze-drying (Chatterjee et al. 2005a). At low glycine to sugar ratios, glycine can be present in the amorphous state and is able to improve protein stability. Being a low-molecular-weight plasticiser, glycine can increase solid-state stability of mAbs despite lowering T_g (Meyer et al. 2009). Amorphous glycine was found to increase the activity of freeze-dried factor V and factor VIII formulations (Hubbard et al. 2007).

Arginine has weak affinity to the protein surface, which indicates that the stabilisation effect of arginine is not due to preferential exclusion. It is able to form hydrogen bonds with protein molecules upon lyophilisation, and in addition ion-dipole interactions are possible owing to its positive charge (Schneider et al. 2011; Arakawa et al. 2007; Ohtake et al. 2011). The mechanisms on the molecular level are still discussed, but protein-protein interactions and aggregation are prevented mainly because of weak binding of arginine to the proteins (Arakawa et al. 2007).

In mixtures with hydroxy di- or tricarboxylic acids, arginine and histidine show bell-shaped changes in the T_g' profiles, with low T_g' values for low and high carboxylic acid concentrations. For example, L-arginine and pure citric acid have T_g' values of $-44°C$ and $-55°C$ respectively, but in combination at a 1:1 ratio for pH adjustment the T_g' is increased up to a maximum of around $-25°C$ (Izutsu et al. 2009a). The T_g' of histidine increases from $-47°C$ at pH 4.0 to $-31°C$ upon pH adjustment to 7.7 with hydrochloric acid. Addition of sucrose to the histidine-containing formulation leads to an additional increase in T_g' (Österberg and Wadsten 1999). Stärtzel et al. (2015) freeze dried 50 mg/ml mAb in ratios 16:64 and 64:16 arginine to sucrose, testing chloride, citrate, phosphate, and succinate as counter ions. Major cake defects were observed with chloride, whereas all other formulations rendered only minor defects. The T_g values were between $58°C$ and $86°C$ (Stärtzel et al. 2015). Addition of arginine hydrochloride to sugar-based BSA lyophilisates reduced protein aggregation during manufacturing and upon storage (Hackl et al. 2018). Arginine exhibits a stabilising effect also if the product is collapsed, as shown for the aggregation propensity of an IgG1 in arginine phosphate pH 7.3 (Schersch et al. 2010). Albumin used to stabilise recombinant factor VIII during lyophilisation (see section 3.7.1) can be replaced by L-arginine, L-glutamic acid, and L-isoleucine. The mixture of 36 mM arginine, 57 mM glutamic acid, and 7 mM isoleucine was as effective as albumin (Paik et al. 2012).

Adding histidine to sucrose formulations resulted in decreased aggregation levels of an IgG but a perturbed secondary structure. Solid-state hydrogen-deuterium exchange and Fourier-transform infrared spectroscopy did not predict protection through histidine, and ionic interactions or suppressed dynamics might be the reason for the perturbation (Moussa et al. 2018). Freeze-drying LDH in pure histidine buffer (10–150 mM, pH 7.3), which resulted in increased enzyme activity with higher histidine concentration, was found optimal at pH 6, higher than citrate and phosphate buffers (Al-Hussein and Gieseler 2013). Both for fibroblast growth factor 21 and catalase, testing different formulations demonstrated stabilising effects of various amino acids. Regarding aggregation and bioactivity of fibroblast growth factor 21, a formulation containing 2% mannitol, 2% trehalose, 0.05% glycine, and 0.1% Poloxamer 188 preserved the protein best (Yang et al. 2018). Alanine, glycine, serine, arginine, histidine, lysine, 4-hydroxy proline, and threonine showed a stabilising effect on catalase activity, while the stabilising effect of other amino acids was weaker and concentration dependent (Lale et al. 2011).

Phenylalanine, isoleucine, and leucine crystallise during lyophilisation and can therefore be used as bulking agents. Starting at low ratios of 2.5:47.5 amino acid to sucrose, the bulking efficiency improved, and a pharmaceutical elegant cake was obtained at a ratio of 5:45. Neither 2 mg/ml nor 50 mg/ml mAb suppressed crystallization of leucine and isoleucine (Horn et al. 2018b). Thus, compared to mannitol and glycine, less amino acid is required to form a crystalline scaffold. Consequently, more sugar can be used in the formulation to generate a higher stabiliser to protein ratio.

L-Glutathione can be used as antioxidant to increase the storage stability of proteins, for example, factor VIII (Jameel et al. 2009). An antioxidative effect can also be achieved with methionine already at rather low concentrations (Kamerzell et al. 2011; Lipiainen et al. 2015; Gervasi et al. 2018). Exemplarily, some commercial hormone (Pergoveris®, Gonal-f®) and coagulation factor (NovoEight®, NovoSeven®) products contain methionine as antioxidant.

Glycylglycine is used as buffering agent for stabilisation of freeze-dried factor VIIa (NovoSeven®), which showed, along with glycine, stabilising effects during freeze-thaw experiments (Enomoto 1989), but a specific stabilising effect in lyophilisation has not been reported yet (Lim et al. 2016; Jensen et al. 2012).

3.4 POLYMERS

Dry amorphous polymer matrices may stabilise proteins similarly to sugar matrices (see also 3.2.3). They were thought to even be superior owing to their high T_g values, as already described above in sugar polymers (see 3.2.3). However, they are less capable of water replacement, and therefore combinations with small-molecule excipients are typically required to achieve adequate protein stability. Polymers may also prevent proteins from aggregating owing to their surface active properties, through sterical hindrance of protein-protein interactions as well as via increased viscosity, reducing protein structural movement (Wang 2005; Wang et al. 2010b).

However, polymers may trigger phase separation, which can have a detrimental effect on protein stability (Ó'Fágáin and Colliton 2017).

Adding up to 20% PEG decreased the T_g' of 10% sucrose to −48°C. Upon annealing at −25°C, PEG crystallised (Bhatnagar et al. 2010). Higher PEG concentrations show the risk of phase separation, and the protein may concentrate in one phase, increasing the risk of aggregation (Ó'Fágáin and Colliton 2017; Schein 1990). Initially amorphous PEG is prone to crystallise during primary drying, which could induce protein aggregation. Interestingly, this crystallization could be inhibited by higher sucrose amounts (Bhatnagar et al. 2011). A collapsed lyophilisate containing LDH with sucrose or trehalose to PEG at a ratio of 3:2 showed the same monomer content and subvisible particle level as a noncollapsed cake (Schersch et al. 2010). On long-term stability at 40°C for 26 weeks, the collapsed cakes remained the same or showed higher monomer content and less subvisible particle level as noncollapsed cakes (Schersch et al. 2012). Interestingly, trehalose glycopolymers (2–50 kDa) resulted in higher remaining activity than polyethylene glycol after freeze-drying and thermal stress (Maynard et al. 2018). Adding copovidone, a copolymer of 1-vinl-2-pyrrolidone (60%) and vinyl acetate (40%) to a lysozyme formulation without further excipients maintained biological activity and conformation integrity during freeze-drying and subsequent storage (Haj-Ahmad et al. 2016). Similar results were found for PVA on the stability of β-galactosidase (Yoshioka et al. 2000).

3.5 SURFACTANTS

During freezing, proteins are exposed to the ice-water interface, which can contribute to freezing-induced protein denaturation and aggregation. By addition of surfactants, proteins can be protected from both freezing- and surface-induced denaturation (Chang et al. 1996). Surfactants may also potentially foster protein refolding, reducing aggregation in the reconstitution step (Carpenter et al. 2002; Jones et al. 2001). The most commonly used surfactants in commercial lyophilized protein products are polysorbate 20 and 80. Exemplarily, polysorbate 80 is used for Benlysta®, Inflectra®/Flixabi®, and Empliciti®, and polysorbate 20 for Xolair® and Herceptin® products (Gervasi et al. 2018). Polysorbate is prone to degradation, such as oxidation, autoxidation, and hydrolysis, which are influenced by various factors including pH, temperature, and oxygen level. Degradation is more pronounced in liquid than in lyophilised state (Hall et al. 2016; Martos et al. 2017; Ha et al. 2002).

Poloxamer 188 is a water-soluble surface-active polyoxyethylene-polyoxypropylene triblock copolymer. It can also compete with the protein for interfaces, preventing adsorption-induced conformational changes and aggregation (Ohtake et al. 2011), and is exemplarily used for Gazyvaro®.

Trehalolipids are reported to have surfactant-like abilities (Franzetti et al. 2010; Schiefelbein 2011). They are of interest as stabilisers for lyophilisation processes owing to their glass-forming ability (Ogawa and Osanai 2012). Sugar-based surfactants can be used as alternative to polysorbate eliminating the instability-prone polyoxyethylene moiety (Schiefelbein 2011). 6-O-octanoyl trehalose and 6-O-lauroyl raffinose, illustrated in Figure 3.2, showed a better efficiency in maintaining LDH activity than pure trehalose. The T_g of both sugar-based lipid formulations was

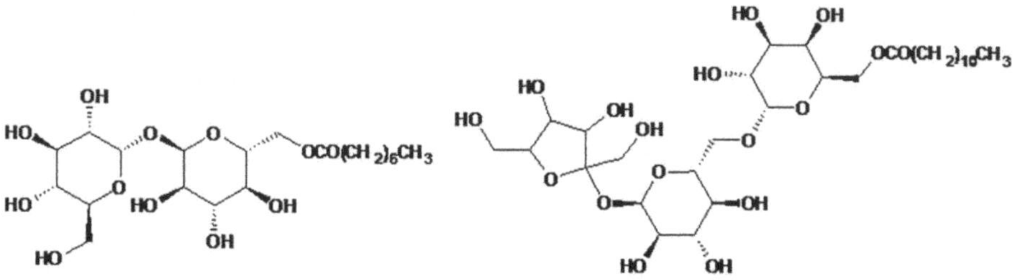

6-O-Octanoyl trehalose   6-O-Lauroyl raffinose

FIGURE 3.2   Structures 6-O-octanoyl trehalose and 6-O-lauroyl raffinose.

approximately 85°C (Ogawa et al. 2016). Trehalolipids did prevent aggregation of IL-11, similarly to polysorbate 80, during freeze-drying and upon short-term storage at 50°C for 1 week (Schiefelbein 2011).

3.6   BUFFER

Protein injectables typically require pH adjustment and stabilisation with buffers. Exemplary well-known buffer systems are histidine (see amino acids in buffer systems in section 3.3) and phosphate, succinate, or citrate (Zbacnik et al. 2017). Buffers used in commercially available freeze-dried antibody drugs are shown in Table 3.1. Some buffer species can selectively crystallise, causing pH changes. Disodium phosphate has lower solubility than monosodium phosphate, potentially leading to earlier crystallization upon concentration during freezing and causing a pH shift of up to 3 pH units (Anchordoquy and Carpenter 1996; Pikal-Cleland et al. 2002; Sarciaux et al. 1999; Bhatnagar et al. 2007). This effect is less pronounced for potassium phosphate buffers (Hellerbrand et al. 2001). Furthermore, succinate buffer is prone to crystallization and therefore to pH shifts, which can be inhibited by amorphous sucrose, trehalose, glycine, and mannitol (Sundaramurthi and Suryanarayanan 2011). Higher antibody concentrations show self-buffering characteristics as well as inhibition of the beforementioned pH shifts. At 50 mg/ml, antibody can act similar to 6 mM citrate or 14 mM histidine buffer (Karow et al. 2013).

Buffers not only affect protein stability via the pH but also by potential direct interaction or changing the mobility in the glassy matrix. During freezing protein stability can be increased by phosphate and citrate, which do not act as strong hydrogen-bond formers, in formulations with stabilisers such as sucrose, which is preferentially excluded in solution (Arsiccio and Pisano 2018). It is often proposed to use low buffer salt concentrations owing to their ability to act as a plasticiser in sugar-based systems and thereby diminish protein stability. Since buffer salts as well as sodium chloride added as tonicity agent or resulting from pH-adjustment with acid or base decrease the T_g' of the formulations, the collapse tendency is increased (Goshima et al. 2016). However, the dibasic phosphate ion HPO_4^{2-} is also able to increase the T_g of sugar formulations at higher pH values. While trehalose showed a T_g of 94°C at different pH values, the T_g was increased by phosphate to

TABLE 3.1
Commonly Used Buffers for Freeze-Dried Antibody Drugs Commercially Available, Adapted from Gervasi et al. (2018)

Buffer	International Nonproprietary Name (INN)	Trade Name
Citrate	belimumab	Benlysta®
	blinatumomab	Blincyto®
	brentuximab vedotin	Adcetris®
	etoluzumab	Empliciti®
Histidine	canakinumab	Cosentyx®
	omalizumab	Xolair®
	pembrolizumab	Keytruda®
	secukinumab	Cosentyx®
	siltuximab	Sylvant®
	trastuzumab	Herceptin®; Herzuma®; Kanjinti®; Ontruzant®
	vedolizumab	Entyvio®
Lysine	blinatumomab	Blicyto®
Phosphate	basiliximab	Simulect®
	infliximab	Flixabi®; Inflectra®; Remicade®; Remsima®
	gemtuzumab ozogamicin	Mylotarg®
	mepolizumab	Nucala®
Succinate	trastuzumab emtansine	Kadcyla®
TRIS	inotuzumab ozogamicin	Besponsa®

119°C at pH 7.5 (Ohtake et al. 2004). Through self-aggregation of HPO_4^{2-} ions, a hydrogen network with trehalose is built, whereas the monobasic phosphate ion $H_2PO_4^-$ does not form such a network and decreases the T_g, acting as a plasticiser (Weng and Elliott 2015). Carboxylic acids and their sodium salts, such as sodium citrate, protected the secondary structure of BSA and IgG better with increasing concentrations. This suggested stabilisation through direct interactions by substitution of water molecules (Izutsu et al. 2009b). Furthermore, addition of sodium citrate to a sucrose formulation led to an increase in T_g from 70°C up to 150°C at a sodium citrate to sucrose ratio of 3:1 (Kets et al. 2004). The stability of highly concentrated mAb (40–160 mg/ml) freeze-dried with sucrose or trehalose was substantially improved in the presence of succinate buffer, as compared to a buffer-free system (Garidel et al. 2015). Furthermore, sodium tetraborate increased T_g' and T_g of BSA/ sugar formulations (Izutsu et al. 2004). The buffering agent dimethyl-succinate was found to stabilise lignin peroxidase as effectively as sucrose and enabled fast primary drying (Capolongo et al. 2003).

3.7 OTHERS

3.7.1 PROTEINS

HSA, also in the recombinant form, can prevent protein adsorption to surfaces and act as water replacement molecules, maintaining protein activity (Hawe 2006; Anchordoquy and Carpenter 1996). HSA is used mostly for solubilising and protecting low-dosed and hydrophobic proteins such as cytokines, coagulation factors, or botulinum toxin (Lipiainen et al. 2015; Hawe 2006). The utilisation of HSA in parenteral drug products comes along with immunogenicity concerns because of the proteinaceous nature of the excipient and the potential formation of mixed aggregates. Additionally, HSA impacts protein analytics. Replacement of HSA in lyophilised products can be achieved by combining surfactants with an amino acid like glycine, arginine, or histidine (Hawe 2006; Haselbeck 2003; Österberg et al. 1997).

Other protein excipients that are not approved for use in human parenteral drug products are intrinsically disorders proteins (IDPs). IDPs can act as cryoprotectants, and an IDP to LDH ratio of 10:1 resulted in similar activity levels after freeze-drying as BSA (Matsuo et al. 2018). Comparing different proteinaceous additives, only the weakly acidic proteins BSA and ovalbumin stabilised LDH as well as other enzymes; however, the effectiveness was limited as compared to sugars (Shimizu et al. 2017).

3.7.2 ALCOHOLS

In the past, cosolvent systems have been used to improve dissolution characteristics and drying times of lyophilised products (Teagarden and Baker 2002; Kunz et al. 2018). The addition of 0.5% and more TBA to a albiglutide (GLP-1 agonist) formulation resulted in a reduction of reconstitution time (Kranz and Rinella 2014).

3.7.3 PRESERVATIVES

Multidose protein formulations require preservation. Intron A® (Interferon α-2b) is a commercially available multidose product that contains benzyl alcohol as preservative in the solid. In other cases, benzyl alcohol is part of the reconstitution medium, for example, in Novarel®, Pregnyl®, and Profasi®, all three containing human chorionic gonadotropin, and Nutropic®, containing human growth hormone (Lim et al. 2016). Benzyl alcohol can substantially decrease the protein aggregation temperature and can increase protein aggregation in liquid (Bis et al. 2015). Reconstitution of rhIL-1ra with 0.9% benzyl alcohol in water resulted in a higher level of aggregation than reconstitution with water for injection. Nevertheless, once in a liquid state, benzyl alcohol did not accelerate aggregation of rhIL-1ra at room temperature (Roy et al. 2005). CroFab® (Crotalidae polyvalent immune Fab) contains thiomersal as preservative in the lyophilisate (Lim et al. 2016).

3.8 EXPERT OPINION

Excipients are crucial for providing the chemical and physical stability of protein drugs in lyophilisates. The excipients serve different functions and sometimes combine several roles. A substantial list of protein-stabilising excipients has been identified. However, in most cases formulators stick to the good old approved sucrose, or potentially trehalose, combine it with surfactant and buffer selected from a small portfolio, if necessary, complement this with crystalline-bulking agent, typically mannitol, and specific stabilisers, which usually function both in the liquid and the dry state.

Sucrose is the gold standard for providing an amorphous matrix and water replacement. Only a few sugars and amino acids are good alternatives, specifically trehalose, HP-β-CD, or arginine. Bulking agents provide pharmaceutically elegant cakes and potentially enable faster primary drying. Mannitol plays the leading role, with glycine, phenylalanine, leucine, and isoleucine as potential other options. Cake elegance also comes with high protein drug content and increased total solid content. Overall, the need for new bulking agents is low, except for bulking agents that can be used at lower concentration in order to keep the sugar stabiliser content high considering isotonicity. In this context reconstitution time, which can become very long at high protein drug concentration, may need to be considered upon excipient selection, as total solid content and cake structure affect the water penetration into the lyophilized cake.

Some excipients under evaluation do not lead to improved protein stability. Instead they provide higher T_g' for more robust primary drying; higher T_g is not necessarily correlated with stability and lyophilisate elegance. A few larger saccharides and polysaccharides like melezitose, raffinose, inulin, dextran, maltodextrin, HEC, and HES serve this function.

The most suitable pH is often identified already in a liquid screening. The buffer should be selected considering its crystallization tendency and potential pH shift upon freezing, which can harm the protein and may lead to a discrepancy between the pH value adjusted in the liquid state and the theoretical pH in the solid state.

In most cases, surfactant is added, again selected from a small group consisting of polysorbate 20, polysorbate 80, and poloxamer 188, although many more surface-active molecules would be available. Addition of an antioxidant is advisable in the case of highly oxidation-sensitive protein molecules. The need for a preservative is rare and it may also be a part of the reconstitution medium.

REFERENCES

Al-Hussein, A., and H. Gieseler. 2013. Investigation of histidine stabilizing eon LDH during freeze-drying. *J. Pharm. Sci.* 102:813–826.

Al-Hussein, A., and H. Gieseler. 2015. Investigation of the stabilizing effects of hydroxyethyl cellulose on LDH during freeze drying and freeze thawing cycles. *Pharm. Dev. Technol.* 20:50–59.

Allison, S. D., Manning, M. C., Randolph, T. W., Middleton, K., Davis, A., and J. F. Carpenter. 2000. Optimization of storage stability of lyophilized actin using combinations of disaccharides and dextran. *J. Pharm. Sci.* 89:199–214.

Anchordoquy, T. J., and J. F. Carpenter. 1996. Polymers protect lactate dehydrogenase during freeze-drying by inhibiting dissociation in the frozen state. *Arch. Biochem. Biophys.* 332:231–238.

Arakawa, T., Tsumoto, K., Kita, Y., Chang, B., and D. Ejima. 2007. Biotechnology applications of amino acids in protein purification and formulations. *Amino Acids* 33:587–605.

Arsiccio, A., and R. Pisano. 2018. The preservation of lyophilized human growth hormone activity: How do buffers and sugars interact? *Pharm. Res.* 35:131–143.

Bhatnagar, B. S., Bogner, R. H., and M. J. Pikal. 2007. Protein stability during freezing: Separation of stresses and mechanisms of protein stabilization. *Pharm. Dev. Technol.* 12:505–523.

Bhatnagar, B. S., Martin, S. M., Teagarden, D. L., Shalaev, E. Y., and R. Suryanarayanan. 2010. Investigation of PEG crystallization in frozen PEG-sucrose-water solutions. I. Characterization of the nonequilibrium behavior during freeze-thawing. *J. Pharm. Sci.* 99:2609–2619.

Bhatnagar, B. S., Martin, S. W. H., Hodge, T. S., et al. 2011. Investigation of PEG crystallization in frozen and freeze-dried PEGylated recombinant human growth hormone – sucrose systems: Implications on storage stability. *J. Pharm. Sci.* 100:3062–3075.

Bis, R.,L., Singh, S. M., Cabello-Villegas, J., and K. M. Mallela. 2015. Role of benzyl alcohol in the unfolding and aggregation of interferon alpha-2a. *J. Pharm. Sci.* 104:407–415.

Bosch, T. 2014. Aggressive freeze-drying – a fast and suitable method to stabilize biopharmaceuticals. PhD diss., Ludwig-Maximilians-Universität München.

Capolongo, A., Barresi, A. A., and G. Rovero. 2003. Freeze-drying of lignin peroxidase: Influence of lyoprotectants on enzyme activity and stability. *J. Chem. Technol. Biotechnol.* 78:56–63.

Carpenter, J. F., Chang, B. S., Garzon-Rodriguez, W., and T. W. Randolph. 2002. Rational design of stable lyophilized protein formulations: Theory and practice. In *Rationale design of stable protein formulations – Theory and practice*, ed. J. F. Carpenter, and M. C. Manning. New York: Kluwer Academic/Plenum publishers.

Cavatur, R. K., Vemuri, N. M., Pyne, A., Chrzan, Z., Toledo-Velasquez, D., and R. Suryanarayanan. 2002. Crystallization behavior of mannitol in frozen aqueous solutions. *Pharm. Res.* 19:894–900.

Chang, B. S., and S. Y. Patro. 2004. Freeze-drying process development for protein pharmaceuticals. In *Lyophilization of biopharmaceuticals*, ed. H. R. Costantino, and M. J. Pikal, 113–138. Arlington: American Association of Pharmaceutical Scientists.

Chang, B. S., Kendrick, B. S., and J. F. Carpenter. 1996. Surface-induced denaturation of proteins during freezing and its inhibition by surfactants. *J. Pharm. Sci.* 85:1325–1330.

Chang, L. L., Shepherd, D., Sun, J., Tang, X. C., and M. J. Pikal. 2005. Effect of sorbitol and residual moisture on the stability of lyophilized antibodies: Implications for the mechanism of protein stabilization in the solid state. *J. Pharm. Sci.* 94:1445–1455.

Chatterjee, K., Shalaev, E. Y., and R. Suryanarayanan. 2005a. Partially crystalline systems in lyophilization: II. Withstanding collapse at high primary drying temperatures and impact on protein activity recovery. *J. Pharm. Sci.* 94:809–820.

Chatterjee, K., Shalaev, E. Y., and R. Suryanarayanan. 2005b. Raffinose crystallization during freeze-drying and its impact on recovery of protein activity. *Pharm. Res.* 22:303–309.

Cicerone, M. T., and C. L. Soles. 2004. Fast dynamics and stabilization of proteins: Binary glasses of trehalose and glycerol. *Biophys. J.* 86:3836–3845.

Cicerone, M. T., Pikal, M. J., and K. K. Qian. 2015. Stabilization of proteins in solid form. *Adv. Drug Del. Rev.* 93:14–24.

Davidson, P., and W. Q. Sun. 2001. Effect of sucrose/raffinose mass ratios on the stability of co-lyophilized protein during storage above the Tg. *Pharm. Res.* 18:474–479.

Davis, J. M., Zhang, N., Payne, R. W., et al. 2013. Stability of lyophilized sucrose formulations of an IgG1: Subvisible particle formation. *Pharm. Dev. Technol.* 18:883–896.

Enomoto, M. 1989. Method for stabilizing blood coagulation factors. European Patent 0359201B1 filed September 12, 1989.

Faghihi, H., Khalili, F., Amini, M., and A. Vatanara. 2017. The effect of freeze-dried antibody concentrations on its stability in the presence of trehalose and hydroxypropyl-beta-cyclodextrin: A Box-Behnken statistical design. *Pharm. Dev. Technol.* 22:724–732.

Faghihi, H., Merrikhihaghi, S., Ruholamini Najafabadi, A., Ramezani, V., Sardari, S., and A. Vatanara. 2016. A comparative study to evaluate the effect of different carbohydrates on the stability of immunoglobulin G during lyophilization and following storage. *Pharm. Sci.* 22:251–259.

Forney-Stevens, K. M., Bogner, R. H., and M. J. Pikal. 2016. Addition of amino acids to further stabilize lyophilized sucrose-based protein formulations: I. Screening of 15 amino acids in two model proteins. *J. Pharm. Sci.* 105:697–704.

Frank, D. W., Gray, J. E., and R. N. Weaver. 1976. Cyclodextrin nephrosis in the rat. *Am. J. Pathol.* 83:367–382.

Franks, F. 1998. Freeze-drying of bioproducts: Putting principles into practice. *Eur. J. Pharm. Biopharm.* 45:221–229.

Franzetti, A., Gandolfi, I., Bestetti, G., Smyth, T. J. P., and I. M. Banat. 2010. Production and applications of trehalose lipid biosurfactants. *Eur. J. Lipid Sci. Technol.* 112:617–627.

Frijlink, H. W., Franssen, E. J. F., Eissens, A. C., Oosting, R., Lerk, C. F., and D. K. F. Meijer. 1991. The effect of parenterally administered cyclodextrins on cholesterol levels in the Rat. *Pharm. Res.* 8:380–384.

Fujii, K., Izutsu, K.-I., Kume, M., et al. 2015. Physical characterization of meso-erythritol as a crystalline bulking agent for freeze-dried formulations. *Chem. Pharm. Bull. (Tokyo)* 63:311–317.

Garidel, P., Pevestorf, B., and S. Bahrenburg. 2015. Stability of buffer-free freeze-dried formulations: A feasibility study of a monoclonal antibody at high protein concentrations. *Eur. J. Pharm. Biopharm.* 97:125–139.

Garzon-Rodriguez, W., Koval, R. L., Chongprasert, S., et al. 2004. Optimizing storage stability of lyophilized recombinant human interleukin-11 with disaccharide/hydroxyethyl starch mixtures. *J. Pharm. Sci.* 93:684–696.

Geidobler, R. M. 2014. Cyclodextrins as excipients in drying of proteins and controlled ice nucleation in freeze-drying. PhD diss., Ludwig-Maximilians-Universität München.

Gervasi, V., Dall'Agnol, R., Cullen, S., McCoy, T., Vucen, S., and A. Crean. 2018. Parenteral protein formulations: An overview of approved products within the European Union. *Eur. J. Pharm. Biopharm.* 131:8–124.

Golovanov, A. P., Hautbergue, G. M., Wilson, S. A., and L.-Y. Lian. 2004. A simple method for improving protein solubility and long-term stability. *J. Am. Chem. Soc.* 126:8933–8939.

Goshima, H., Forney-Stevens, K. M., Liu, M., et al. 2016. Addition of monovalent electrolytes to improve storage stability of freeze-dried protein formulations. *J. Pharm. Sci.* 105:530–541.

Gould, S., and R. C. Scott. 2005. 2-Hydroxypropyl-beta-cyclodextrin (HP-beta-CD): A toxicology review. *Food Chem. Toxicol.* 43:1451–1459.

Ha, E., Wang, W., and Y. J. Wang. 2002. Peroxide formation in polysorbate 80 and protein stability. *J. Pharm. Sci.* 91:2252–2264.

Hackl, E., Darkwah, J., Smith, G., and I. Ermolina. 2018. Effect of arginine on the aggregation of protein in freeze-dried formulations containing sugars and polyol: II. BSA reconstitution and aggregation. *AAPS PharmSciTech* 19:2934–2947.

Haj-Ahmad, R. R., Mamayusupov, M., Elkordy, E. A., and A. A. Elkordy. 2016. Influences of copolymers (Copovidone, Eudragit RL PO and Kollicoat MAE 30 DP) on stability and bioactivity of spray-dried and freeze-dried lysozyme. *Drug Dev. Ind. Pharm.* 42:2086–2096.

Hall, T., Sandefur, S. L., Frye, C. C., Tuley, T. L., and L. Huang. 2016. Polysorbates 20 and 80 degradation by group XV lysosomal phospholipase A2 isomer X1 in monoclonal antibody formulations. *J. Pharm. Sci.* 105:1633–1642.

Hancock, B. C., and G. Zografi. 1997. Characteristics and significance of the amorphous state in pharmaceutical systems. *J. Pharm. Sci.* 86:1–12.

Haselbeck, A. 2003. Epoetins: Differences and their relevance to immunogenicity. *Curr. Med. Res. Opin.* 19:430–432.

Hawe, A. 2006. Studies on stable formulations for a hydrophobic cytokine. PhD diss., Ludwig-Maximilians-Universität München.

Hawe, A., and W. Friess. 2006. Impact of freezing procedure and annealing on the physico-chemical properties and the formation of mannitol hydrate in mannitol – sucrose – NaCl formulations. *Eur. J. Pharm. Biopharm.* 64:316–325.

Heljo, P. 2013. Comparison of disaccharides and polyalcohols as stabilizers in freeze-dried protein formulations. PhD diss., University of Helsinki.

Hellerbrand, K., Papadimitriou, A., and G. Winter. 2001. Process for stabilizing proteins. U.S. Patent 6238664 filed May 29, 2001.

Horn, J., Schanda, J., and W. Friess. 2018a. Impact of fast and conservative freeze-drying on product quality of protein-mannitol-sucrose-glycerol lyophilizates. *Eur. J. Pharm. Biopharm.* 127:342–354.

Horn, J., Tolardo, E., Fissore, D., and W. Friess. 2018b. Crystallizing amino acids as bulking agents in freeze-drying. *Eur. J. Pharm. Biopharm.* 132:70–82.

Hubbard, A., Bevan, S., and P. Matejtschuk. 2007. Impact of residual moisture and formulation on Factor VIII and Factor V recovery in lyophilized plasma reference materials. *Anal. Bioanal. Chem.* 387:2503–2507.

Ivarsson, E., Knutsson, J., Rippner, B., Nilsson, U., and I. Agerkvist. 2013. New stabilizing agent for pharmaceutical proteins. U.S. Patent Application 2013/0116410 A1 filed May 9, 2013.

Ivarsson, E., Knutsson, J., Rippner, B., Nilsson, U., and I. Agerkvist. 2018. Stabilizing agent for pharmaceutical proteins. U.S. Patent 9943600 B2 filed April 17, 2018.

Iwai, J., Ogawa, N., Nagase, H., Endo, T., Loftsson, T., and H. Ueda. 2007. Effects of various cyclodextrins on the stability of freeze-dried lactate dehydrogenase. *J. Pharm. Sci.* 96:3140–3143.

Izutsu, K.-I., Aoyagi, N., and S. Kojima. 2004. Protection of protein secondary structure by saccharides of different molecular weights during freeze-drying. *Chem. Pharm. Bull. (Tokyo)* 52:199–203.

Izutsu, K.-I., Kadoya, S., Yomota, C., Kawanishi, T., Yonemochi, E., and K. Terada. 2009a. Freeze-drying of proteins in glass solids formed by basic amino acids and dicarboxylic acids. *Chem. Pharm. Bull.* 57:43–48.

Izutsu, K.-I., Kadoya, S., Yomota, C., Kawanishi, T., Yonemochi, E., and K. Terada. 2009b. Stabilization of protein structure in freeze-dried amorphous organic acid buffer salts. *Chem. Pharm. Bull. (Tokyo)* 57:1231–1236.

Izutsu, K.-I., Yoshioka, S., and T. Terao. 1994. Effect of mannitol crystallinity on the stabilization of enzymes during freeze-drying. *Chem. Pharm. Bull. (Tokyo)* 42:5–8.

Izutsu, K.-I., Yoshioka, S., Kojima, S., Randolph, T. W., and J. F. Carpenter. 1996. Effects of sugars and polymers on crystallization of poly(ethylene glycol) in frozen solutions: Phase separation between incompatible polymers. *Pharm. Res.* 13:1393–1400.

Izutsu, K. -I., Yoshida, H., Shibata, H., and Y. Goda. 2016. Amorphous-amorphous phase separation of freeze-concentrated protein and amino acid excipients for lyophilized formulations. *Chem. Pharm. Bull. (Tokyo)* 64:1674–1680.

Jameel, F., Tchessalov, S., Bjornson, E., Lu, X., Besman, M., and M. Pikal. 2009. Development of freeze-dried biosynthetic Factor VIII: I. A case study in the optimization of formulation. *Pharm. Dev. Technol.* 14:687–697.

Jensen, M. B., Hansen, B. L., and T. Kornfelt. 2012. Stabilised solid compositions of factor VII polypeptides. U.S. Patent 8299029 B2 filed October 30, 2012.

Jones, L. S., Randolph, T. W., Kohnert, U., et al. 2001. The effects of tween 20 and sucrose on the stability of anti-L-selectin during lyophilization and reconstitution. *J. Pharm. Sci.* 90:1466–1477.

Kadoya, S., Fujii, K., Izutsu, K.-I., et al. 2010. Freeze-drying of proteins with glass-forming oligosaccharide-derived sugar alcohols. *Int. J. Pharm.* 389:107–113.

Kamerzell, T. J., Esfandiary, R., Joshi, S. B., Middaugh, C. R., and D. B. Volkin. 2011. Protein – excipient interactions: Mechanisms and biophysical characterization applied to protein formulation development. *Adv. Drug Del. Rev.* 63:1118–1159.

Karow, A. R., Bahrenburg, S., and P. Garidel. 2013. Buffer capacity of biologics – From buffer salts to buffering by antibodies. *Biotechnol. Prog.* 29:480–492.

Kasraian, K., Spitznagel, T. M., Juneau, J. A., and K. Yim. 1998. Characterization of the sucrose/glycine/water system by differential scanning calorimetry and freeze-drying microscopy. *Pharm. Dev. Technol.* 3:233–239.

Kets, E. P., IJpelaar, P. J., Hoekstra, F. A., and H. Vromans. 2004. Citrate increases glass transition temperature of vitrified sucrose preparations. *Cryobiology* 48:46–54.

Kranz, J., and J. Rinella. 2014. Lyophilized formulations. U.S. Patent 20140044717 A1 filed February 13, 2014.

Kunz, C., Schuldt-Lieb, S., and H. Gieseler. 2018. Freeze-drying from organic cosolvent systems, Part 1: Thermal analysis of cosolvent-based placebo formulations in the frozen state. *J. Pharm. Sci.* 107:887–896.

Lale, S. V., Goyal, M., and A. K. Bansal. 2011. Development of lyophilization cycle and effect of excipients on the stability of catalase during lyophilization. *Int. J. Pharm. Investig.* 1:214–221.

Liebner, R., Bergmann, S., Hey, T., Winter, G., and A. Besheer. 2015. Freeze-drying of HESylated IFNalpha-2b: Effect of HESylation on storage stability in comparison to PEGylation. *Int. J. Pharm.* 495:608–611.

Lim, J. Y., Kim, N. A., Lim, D. G., Kim, K. H., Choi, D. H., and S. H. Jeong. 2016. Process cycle development of freeze drying for therapeutic proteins with stability evaluation. *J. Pharm. Investig.* 46:519–536.

Lipiainen, T., Peltoniemi, M., Sarkhel, S., et al. 2015. Formulation and stability of cytokine therapeutics. *J. Pharm. Sci.* 104:307–326.

Liu, J. 2006. Physical characterization of pharmaceutical formulations in frozen and freeze-dried solid states: Techniques and applications in freeze-drying development. *Pharm. Dev. Technol.* 11:3–28.

Lueckel, B., Bodmer, D., Helk, B., and H. Leuenberger. 1998. Formulations of sugars with amino acids or mannitol – influence of concentration ratio on the properties of the freeze-concentrate and the lyophilizate. *Pharm. Dev. Technol.* 3:325–336.

Martos, A., Koch, W., Jiskoot, W., et al. 2017. Trends on analytical characterization of polysorbates and their degradation products in biopharmaceutical formulations. *J. Pharm. Sci.* 106:1722–1735.

Matsuo, N., Goda, N., Shimizu, K., Fukuchi, S., Ota, M., and H. Hiroaki. 2018. Discovery of cryoprotective activity in human genome-derived intrinsically disordered proteins. *Int. J. Mol. Sci.* 19, 401.

Mattern, M., Winter, G., Kohnert, U., and G. Lee. 1999. Formulation of proteins in vacuum-dried glasses. II. Process and storage stability in sugar-free amino acid systems. *Pharm. Dev. Technol.* 4:199–208.

Maynard, H. D., Mancini, R. J., Lee, J., and E.-W. Lin. 2018. Stabilization of biomolecules using sugar polymers. U.S. Patent 9901648 B2 filed February 27, 2018.

Meister, E., Sasic, S., and H. Gieseler. 2009. Freeze-dry microscopy: Impact of nucleation temperature and excipient concentration on collapse temperature data. *AAPS PharmSciTech* 10:582–588.

Meyer, J. D., Nayar, R., and M. C. Manning. 2009. Impact of bulking agents on the stability of a lyophilized monoclonal antibody. *Eur. J. Pharm. Sci.* 38:29–38.

Mollmann, S. H., Bukrinsky, J. T., Elofsson, U., et al. 2008. The stability of insulin in solid formulations containing melezitose and starch. Effects of processing and excipients. *Drug Dev. Ind. Pharm.* 32:765–778.

Moussa, E. M., Singh, S. K., Kimmel, M., Nema, S., and E. M. Topp. 2018. Probing the conformation of an IgG1 monoclonal antibody in lyophilized solids using solid-state hydrogen-deuterium exchange with mass spectrometric analysis (ssHDX-MS). *Mol. Pharm.* 15:356–368.

O'Brien, J. 1996. Stability of trehalose, sucrose and glucose to nonenzymatic browning in model systems. *J. Food Sci.* 61:679–682.

Ó'Fágáin, C., and K. Colliton. 2017. Storage and lyophilization of pure proteins. In *Protein chromatography: Methods and protocols*, ed. D. Walls and S. T. Loughran, 159–190. New York, NY: Springer.

Ogawa, S., and S. Osanai. 2012. Glass transition behavior of aqueous solution of sugar-based surfactants. In *Supercooling*, ed. P. Wilson, 29–54. Rijeka: InTech.

Ogawa, S., Kawai, R., Koga, M., Asakura, K., Takahashi, I., and S. Osanai. 2016. Oligosaccharide-based surfactant/citric acid buffer system stabilizes lactate dehydrogenase during freeze-drying and storage without the addition of natural sugar. *J. Oleo Sci.* 65:525–532.

Ohtake, S., Kita, Y., and T. Arakawa. 2011. Interactions of formulation excipients with proteins in solution and in the dried state. *Adv. Drug Del. Rev.* 63:1053–1073.

Ohtake, S., Schebor, C., Palecek, S. P., and J. J. de Pablo. 2004. Effect of pH, counter ion, and phosphate concentration on the glass transition temperature of freeze-dried sugar-phosphate mixtures. *Pharm. Res.* 21:1615–1621.

Österberg, T., and T. Wadsten. 1999. Physical state of l-histidine after freeze-drying and long-term storage. *Eur. J. Pharm. Sci.* 8:301–308.

Paik, S. H., Kim, Y. J., Han, S. K., Kim, J. M., Huh, J. W., and Y. I. Park. 2012. Mixture of three amino acids as stabilizers replacing albumin in lyophilization of new third generation recombinant factor VIII GreenGene F. *Biotechnol. Prog.* 28:1517–1525.

Passot, S., Fonseca, F., Alarcon-Lorca, M., Rolland, D., and M. Marin. 2005. Physical characterisation of formulations for the development of two stable freeze-dried proteins during both dried and liquid storage. *Eur. J. Pharm. Biopharm.* 60:335–348.

Pikal-Cleland, K. A., Cleland, J. L., Anchordoquy, T. J., and J. F. Carpenter. 2002. Effect of glycine on pH changes and protein stability during freeze-thawing in phosphate buffer systems. *J. Pharm. Sci.* 91:1969–1979.

Pikal, M. J. 1994. Freeze-drying of proteins. In *Formulation and delivery of proteins and peptides*, ed. J. L. Cleland, and R. Langer, 120–133. Washington, DC: American Chemical Society.

Prestrelski, S. J., Pikal, K. A., and T. Arakawa. 1995. Optimization of lyophilization conditions for recombinant human interleukin-2 by dried-state conformational analysis using fourier-transform infrared spectroscopy. *Pharm. Res.* 12:1250–1259.

Prestrelski, S. J., Tedeschi, N., Arakawa, T., and J. F. Carpenter. 1993. Dehydration-induced conformational transitions in proteins and their inhibition by stabilizers. *Biophys. J.* 65:661–671.

Randolph, T. W. 1997. Phase separation of excipients during lyophilization: Effects on protein stability. *J. Pharm. Sci.* 86:1198–1203.

Remmele, R. L., Krishnan, S., and W. J. Callahan. 2012. Development of stable lyophilized protein drug products. *Curr. Pharm. Biotechnol.* 13:471–496.

Roy, S., Jung, R., Kerwin, B. A., Randolph, T. W., and J. F. Carpenter. 2005. Effects of benzyl alcohol on aggregation of recombinant human interleukin-1-receptor antagonist in reconstituted lyophilized formulations. *J. Pharm. Sci.* 94:382–396.

Santagapita, P. R., Brizuela, L. G., Mazzobre, M. F., et al. 2008. Structure/function relationships of several biopolymers as related to invertase stability in dehydrated systems. *Biomacromolecules* 9:741–747.

Sarciaux, J.-M., Mansour, S., Hageman, M. J., and S. L. Nail. 1999. Effects of buffer composition and processing conditions on aggregation of bovine IgG during freeze-drying. *J. Pharm. Sci.* 88:1354–1361.

Schein, C. H. 1990. Solubility as a function of protein structure and solvent components. *Nat. Biotechnol.* 8:308–317.

Schersch, K., Betz, O., Garidel, P., Muehlau, S., Bassarab, S., and G. Winter. 2010. Systematic investigation of the effect of lyophilizate collapse on pharmaceutically relevant proteins I: Stability after freeze-drying. *J. Pharm. Sci.* 99:2256–2278.

Schersch, K., Betz, O., Garidel, P., Muehlau, S., Bassarab, S., and G. Winter. 2012. Systematic investigation of the effect of lyophilizate collapse on pharmaceutically relevant proteins, part 2: Stability during storage at elevated temperatures. *J. Pharm. Sci.* 101:2288–2306.

Schiefelbein, L. 2011. Sugar-based surfactants for pharmaceutical protein formulations. PhD diss., Ludwig-Maximilians-Universität München.

Schneider, C. P., Shukla, D., and B. L. Trout. 2011. Arginine and the Hofmeister series: The role of ion-ion interactions in protein aggregation suppression. *J. Phys. Chem. B* 115:7447–7458.

Sentko, A., and I. Willibald-Ettle. 2012. Isomalt. In *Sweeteners and sugar alternatives in food technology*, ed. K. O'Donnel, and M. W. Kearsley. Oxford: Wiley-Blackwell.

Serno, T., Geidobler, R., and G. Winter. 2011. Protein stabilization by cyclodextrins in the liquid and dried state. *Adv. Drug. Deliv. Rev.* 63:1086–1106.

Shimizu, T., Korehisa, T., Imanaka, H., Ishida, N., and K. Imamura. 2017. Characteristics of proteinaceous additives in stabilizing enzymes during freeze-thawing and -drying. *Biosci. Biotechnol. Biochem.* 81:687–697.

Shiotani, K., Uehata, K., Irie, T., Uekama, K., Thompson, D. O., and V. J. Stella. 1995. Differential effects of sulfate and sulfobutyl ether of β-cyclodextrin on erythrocyte membranes in vitro. *Pharm. Res.* 12:78–84.

Starciuc, T., Guinet, Y., Paccou, L., and A. Hedoux. 2017. Influence of a small amount of glycerol on the trehalose bioprotective action analyzed in situ during freeze-drying of lyzozyme formulations by micro-Raman spectroscopy. *J. Pharm. Sci.* 106:2988–2997.

Stärtzel, P., Gieseler, H., Gieseler, M., et al. 2015. Freeze drying of l-arginine/sucrose-based protein formulations, Part I: Influence of formulation and arginine counter ion on the critical formulation temperature, product performance and protein stability. *J. Pharm. Sci.* 104:2345–2358.

Sundaramurthi, P., and R. Suryanarayanan. 2010a. Influence of crystallizing and non-crystallizing cosolutes on trehalose crystallization during freeze-drying. *Pharm. Res.* 27:2384–2393.

Sundaramurthi, P., and R. Suryanarayanan. 2010b. Trehalose crystallization during freeze-drying: Implications on lyoprotection. *J. Phys. Chem. Lett.* 1:510–514.

Sundaramurthi, P., and R. Suryanarayanan. 2011. The effect of crystallizing and non-crystallizing cosolutes on succinate buffer crystallization and the consequent pH shift in frozen solutions. *Pharm. Res.* 28:374–385.

Tanaka, K., Takeda, T., and K. Miyajima. 1991. Cryoprotective effect of saccharides on denaturation of catalase by freeze-drying. *Chem. Pharm. Bull.* 39:1091–1094.

te Booy, M. P., de Ruiter, R. A., and A. L. de Meere. 1992. Evaluation of the physical stability of freeze-dried sucrose-containing formulations by differential scanning calorimetry. *Pharm. Res.* 9:109–114.

Teagarden, D. L., and D. S. Baker. 2002. Practical aspects of lyophilization using non-aqueous co-solvent systems. *Eur. J. Pharm. Sci.* 15:115–133.

Tonnis, W. F., Mensink, M. A., de Jager, A., van der Voort Maarschalk, K., Frijlink, H. W., and W. L. Hinrichs. 2015. Size and molecular flexibility of sugars determine the storage stability of freeze-dried proteins. *Mol. Pharm.* 12:684–694.

Tuderman, A. K., Strachan, C. J., and A. M. Juppo. 2018. Isomalt and its diastereomer mixtures as stabilizing excipients with freeze-dried lactate dehydrogenase. *Int. J. Pharm.* 538:287–295.

Walsh, G. 2014. Biopharmaceutical benchmarks 2014. *Nat. Biotechnol.* 32:992–1000.

Wang, W. 2005. Protein aggregation and its inhibition in biopharmaceutics. *Int. J. Pharm.* 289:1–30.

Wang, W., Nema, S., and D. Teagarden. 2010a. Protein aggregation – Pathways and influencing factors. *Int. J. Pharm.* 390:89–99.

Wang, W., Li, N., and S. Speaker. 2010b. External factors affecting protein aggregation. In *Aggregation of therapeutic proteins*, ed. W. Wang, and C. J. Roberts. Hoboken, NJ: John Wiley & Sons.

Wang, S., Zhang, N., Hu, T., et al. 2015. Viscosity-lowering effect of amino acids and salts on highly concentrated solutions of two IgG1 monoclonal antibodies. *Mol. Pharm.* 12:4478–4487.

Weinbuch, D., Cheung, J. K., Ketelaars, J., et al. 2015. Nanoparticle impurities in pharmaceutical-grade sugars and their interference with light scattering-based analysis of protein formulations. *Pharm. Res.* 32:2419–2427.

Weng, L., and G. D. Elliott. 2015. Distinctly different glass transition behaviors of trehalose mixed with Na_2HPO_4 or NaH_2PO_4: Evidence for its molecular origin. *Pharm. Res.* 32:2217–2228.

Yang, X., Hui, Q., Yu, B., et al. 2018. Design and evaluation of lyophilized fibroblast growth factor 21 and its protection against ischemia cerebral injury. *Bioconjug. Chem.* 29:287–295.

Yoshioka, S., and Y. Aso. 2007. Correlations between molecular mobility and chemical stability during storage of amorphous pharmaceuticals. *J. Pharm. Sci.* 96:960–981.

Yoshioka, S., Aso, Y., Kojima, S., and T. Tanimoto. 2000. Effect of polymer excipients on the enzyme activity of lyophilized bilirubin oxidase and β-galactosidase formulations. *Chem. Pharm. Bull.* 48:283–285.

Yu, L., Mishra, D. S., and D. R. Rigsbee. 1998. Determination of the glass properties of D-mannitol using sorbitol as an impurity. *J. Pharm. Sci.* 87:774–777.

Zbacnik, T. J., Holcomb, R. E., Katayama, D. S., et al. 2017. Role of buffers in protein formulations. *J. Pharm. Sci.* 106:713–733.

4 Infrared Imaging and Multivariate Image Analysis (MIA)
A New PAT for Freeze-Drying Monitoring and Control

Domenico Colucci, José Manuel Prats-Montalbán, Alberto Ferrer, and Davide Fissore

CONTENTS

4.1 INFRARED IMAGING

In a vacuum freeze-drying (VFD) process, mass and heat transfer are intimately coupled, as discussed in Chapter 1. For this reason, among the many strategies studied in the past years for monitoring and control the VFD process that will be discussed in detail in Chapter 6, the measurement of the temperature of the product in a well-defined position proved to be one of the most efficient and reliable (Fissore et al. 2018). Coupling this measurement with a mathematical model of the process allows inference of the maximum temperature of the product and the residual amount of ice, and, thus, the ending time of the primary drying phase. Soft sensors, based on the high-gain algorithm (Velardi et al. 2010) and the extended Kalman filter (Velardi et al. 2009; Bosca and Fissore 2011; Bosca et al. 2014), were developed and presented to estimate the heat and mass transfer coefficients once a measurement of the temperature, usually at the bottom of the vial, is provided. Moreover, the use of multiple temperature measurements inside the batch allows accounting for in-batch variability (Bosca et al. 2013).

The main drawback of this approach is the way the temperature measurements are obtained. Usually, thermocouples at lab-scale, or the resistance thermal detectors (RTDs) preferred in industrial applications since they are more robust and can be sterilized, are inserted inside the vial, which places the tip in direct contact with the vial's bottom (Willemer 1991; Oetjen and Haseley 2004). Unfortunately, the inclusion of an external object inside the product enhances the nucleation kinetics that will occur at higher temperature, with respect to the other vials of the batch. The effect of this early nucleation is a lower number of bigger ice crystals, that is, a lower resistance to mass transfer inside the dried cake of the product and a lower drying time. The sample where the probe is inserted will not be representative of the whole batch. To cope with this problem, Grassini et al. (2013) presented an innovative system made of a submicrometric film sputtered directly on the vial wall. Furthermore, the necessity of wires and batteries for electrical supply does not fit the requirements of the pharmaceutical industries in terms of sterility and automatization of the process. Wireless systems have been introduced in the recent years (Schneid and Gieseler 2008; Corbellini et al. 2010).

Infrared imaging was proposed in past years as an effective solution to both problems since it allows a noncontact measurement based on the IR radiation emitted by every physical body with a temperature above 0 K, and it is possible to obtain a full and highly accurate map of the temperature distribution of the body also in harsh environmental conditions.

The measurement of the intensity of the electromagnetic radiation that a body emits is called *radiometry* and is usually associated with the measurement of its temperature. The word *infrared* describes the electromagnetic radiation in the wavelength, ranging from visible light, whose upper limit is about 700 nm, to the microwave band, whose lower bound is usually assumed to be 1 mm. Radiometry is almost entirely concerned with the lower end of this spectrum, usually below 15 μm. The radiation commonly detected by IR imaging systems is in the range of 7.5 μm to 15 μm.

The total incident radiation for a given body can be divided into three terms. Part of the radiation is absorbed by the object and will modify its energetical status at atomic level; some will be reflected, and the remaining, if the material of the object allows it, is transmitted through it. The ratio of the absorbed, reflected, and transmitted radiation over the total incident radiation defines respectively the absorbance, α, the reflectance, ρ, and the transmittance, τ. Their sum, for a defined body, must be one. Another factor that must be considered is the emissivity, ε, which represents the amount of energy emitted from an object at a well-defined temperature with respect to the one emitted from a black body at the same temperature. A *black body* is an object that perfectly absorb and emits radiation, that is, according to Kirchoff's law, $\alpha = \varepsilon = 1$. The emissivity, by definition, ranges from 0 to 1 and depends basically on the material and its superficial structure; it changes also with the wavelength of the radiation and the body temperature and might also be influenced by the angle between the sensor and the monitored object.

Planck's law states that the radiance of a black body depends only on its temperature, and thus, we can relate the total radiation emitted from a black body to its actual temperature. In general, for real objects, that is, those having emissivity lower than one, the amount of emitted radiation must be approximated. Usually this is done assuming that it is equal to that of a black body multiplied by its emissivity. This assumption holds true at high emissivity; the lower the emissivity, the less reliable the measured temperature will be. If the emissivity is assumed to be constant across the whole waveband, the body will be called a *gray body*.

An infrared camera basically acts as a transducer, converting the electromagnetic incident radiation into a physical property that could be easily measured, generally an electrical current or a voltage difference. The main element of these sensors is the detector, the actual sensing part. Basically, two kind of detectors exist: thermal and quantum detectors. In a quantum detector the single photon hits a surface made of a semiconductor material, is absorbed, and forms an electron hole pair that will modify the electrical properties of the system. Quantum systems could be either photoconductors or photovoltaics. In the former the conductivity of the element is increased and can be measured by passing a known current through the detector; in the latter the electronic transition creates a voltage difference that is amplified and directly measured. The absorption of the photons will increase the temperature of the sensing elements, leading to the formation of additional electron hole pairs and an error in the current or voltage measurement, the so called *dark current*. The efficiency of the detector will decrease as the dark current increases, becoming zero if the sensing element is completely saturated. To prevent this phenomenon, and keep a certain accuracy in the measurement, these detectors must be cooled at a temperature generally depending on the set up and the required accuracy, but usually very low, for example, 70 K. The cooling is usually achieved with liquid nitrogen (Joule-Thompson cooler), a gas cooler (Stirling cycle), or via the thermoelectrical Peltier effect. Quantum detectors are faster, more accurate, and more sensible than thermal detectors, but the requirement of such a low working temperature makes them unsuitable for large-scale production and applications. Uncooled IR cameras use thermal detectors (pyrometers, microbolometers, etc.) to estimate the incident radiation. When the electromagnetic radiation hits the sensing element, its temperature will increase according to the amount of incident radiation that reaches its surface, as described by the Stefan-Boltzmann law. If we can measure the temperature of this element we obtain an estimation of the radiation absorbed and, thus, a thermal infrared detector. The temperature is inferred by measuring the current flow reduction due to the temperature increment. The dissipation and high thermal capacity of the sensing element have been the main limitations of this technology, overcome by the improvement in the manufacturing of semiconductor elements and of micromachining. Because they are much cheaper than quantum detectors, microbolometers opened the way to the use of thermal imaging technology into all those applications where a trade-off between the costs of the equipment and the quality of the measurement is possible (Pillans 2008; Meola 2012).

4.2 PROCESS MONITORING

Sir John Herschel is generally acknowledged as the father of thermography, since he produced the first "heat picture" back in the 19th century. The main properties of the infrared radiation are to be detectable in low-light conditions and to be only mildly absorbed by air humidity; thus, the main interest into this technology has always been for military applications, namely night watch, target acquisition, homing, and tracking. The first medical and civil applications of the infrared technology appeared in the late 1960s. After the infrared technology was declassified by the U.S. government in 1992, the applications grew exponentially, together with efforts in developing new sensors and improving the existing. A review of the medical application of IR thermography was presented by Lahiri et al. (2012), while Bagavathiappan et al. (2013) presented a review of the application to structural monitoring of civil industries, electrical and electronic devices, equipment deformation and inspection, as well as corrosion and weld monitoring. Samanian and Mohebbi (2016) reviewed the application to the food science industry, including monitoring and defect detection. Some applications of thermography to the monitoring of drying processes of foodstuff were presented by Fito et al. (2004), Albanese et al. (2013), and Cuibus (2013).

Concerning the monitoring of pharmaceutical products, near visible infrared, also called near infrared spectroscopy (NIR), was the first technology to be studied and tested (Brulls et al. 2003). The proximity of these radiations to the visible spectra makes the technology for its detection very cheap compared to the one used for the reading of the remaining part of the infrared spectrum (Pillans 2008). Changes in the physical state and the amount of water inside the vial generate significant changes also in the spectra emitted. Multivariate statistical techniques are usually required to analyze the huge amount of data obtained (De Beer et al. 2009).

A U.S. patent is the first document presenting an imaging system, an RGB camera, that applied the real-time monitoring of the VFD process (Böttger and Häuptle 2009). Emteborg et al. (2014) firstly studied the use of an infrared camera to monitor the evolution of a vacuum freeze-drying process of pharmaceuticals. In their work the camera was mounted on the ceiling of a freeze-dryer; a germanium window, a material mostly transparent to infrared radiation, guarantees a free view inside the chamber. In this configuration the camera is protected from the low temperature, low pressure, and the high moisture level inside the drying chamber, but the camera can only look at the first shelf from the top and monitors freeze-drying processes carried out in bulk or in vials without stoppers. Furthermore, the existing equipment should undergo severe modifications. Also, Van Bockstal et al. (2018) placed the camera outside the drying chamber and used the online thermal measurements to monitor and optimize a continuous spin freeze-drying process. Zane (2010) patented the use of an IR imaging system for monitoring of a pharmaceutical manufacturing process, including the lyophilization step. They considered the possibility of placing the system inside the chamber. Lietta et al. (2019) presented and validated a sensor that could be placed inside the drying chamber. All the electronics, including the infrared camera, an RGB camera, the processing board, and the WiFi antenna for data communication, are enclosed into a case designed to protect the electronics from the surroundings. This enables the use of this sensor in existing equipment, without further modifications, and in any shelf and position inside the chamber. Additionally, they also proved that the sensor shields the vial in front of it from the effects of radiation from the chamber wall, providing, for these vials, a measurement that is representative of the majority of the batch. This system was also used for risk assessment, to evaluate possible deviations in measurements of inserted thermocouples and its effect on process monitoring (Demichela et al. 2018).

4.2.1 Product Temperature

The first information that an IR imaging system provides is of course the information on the temperature of the product, in addition to that of the objects surrounding the vials, that is, the drying chamber. Figure 4.1 shows the evolution of the temperature at the bottom center of ten vials during the whole freeze-drying cycle extracted directly from the thermal images recorded.

All the features that characterize the evolution of the product during the three stages of the process, namely cooling rate, nucleation, crystal growth, primary drying onset and offset, as well as the temperature increment during the secondary drying, are clearly visible. Nevertheless, to obtain a good and reliable measurement of temperature, even on a single spot, is not straightforward.

A radiometric measurement is dependent on many factors, the most important being the correct estimation of the emissivity of the object we are going to monitor. Since the reflective properties of the materials change after a certain angle, measurements should be performed only when the angle between the surface and the sensing element is lower than 45°. Other aspects that must be carefully accounted for are the reflected apparent temperature, the temperature and humidity of the atmosphere inside the chamber, and the distance from the object. Lietta et al. (2019) proved that the values of humidity and temperature inside the drying chamber do not affect the quality of the measurement by more than 0.01°C, and just an average value of the typical conditions occurring during a drying cycle are enough. Given the low pressure inside the chamber and the typical size of a freeze-dryer chamber, the distance from the object is not important in this specific application. The reflected apparent temperature, that is, the set of radiations that the surrounding environment emits toward the measured item and that the same "reflects" toward the thermal camera, together with its own radiation, on the other hand, plays a dramatic role on the

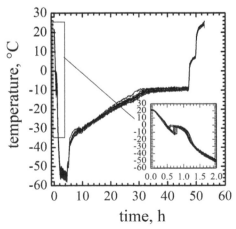

FIGURE 4.1 Raw thermal profile of ten vials, measured at the bottom center, during a VFD cycle. Inner box, details of the freezing step.

quality of the measurement and should be carefully accounted for. The ISO 18434–1 guideline carefully describes the procedures and the precautions required to obtain a good measurement.

Van Bockstal et al. (2018) studied the variability of the transmittance of the germanium window with temperature. Since the radiation emitted by the vial changes with temperature, and the transmittance of the germanium disk varies with the wavelength of the incident radiation, the effective spectral transmittance appears to increase from 0.825 to 0.85 when the temperature of the outer wall of the vial increases from −38°C to −24°C, according to a second order equation. This mild effect on the primary drying might have a dramatic importance if the technology was to be applied to the online monitoring of the freezing stage, which usually occurs at lower temperature.

Another big issue is the low emissivity, from 0.2 to 0.3, of the interiors of the drying chambers, usually made of stainless polished steel. This low emissivity gives back wrong measurements and even more important makes the stainless steel an almost perfect IR mirror that reflects the temperature of the surroundings, including the lens of the camera itself. Although many commercially available systems have a built-in software for emissivity correction, the direct measurement of the temperature of a shelf or of the chamber wall, such as to characterize the radiation on the vials placed at the side of the shelf and to account for the heterogeneity of heating policy inside the chamber, requires covering the region of interest with specific paints or tape at high and known emissivity. Regarding the product, in the case of bulk freeze-drying, most of the solutions of biological material show an emissivity in the range of 0.85 to 0.95. When the product is freeze-dried directly into single-dose units, the glass of the vials becomes our main concern. A value of 0.9 has been measured from Lietta et al. (2019) following the procedure presented in the ISO 18434–1 guideline (Part 1, Annex A.2), while Van Bockstal et al. (2018) reported a value of 0.93. In all cases we are not reading directly the product temperature, but the temperature of the vial glass. Figure 4.2 shows a comparison of the temperature measured from a thermocouple stuck inside the product and the evolution read on five different pixels at the bottom of a single vial.

When comparing the temperature measured by a thermocouple and that obtained using the IR camera, it has to be taken into account that there is a certain uncertainty on the position of the thermocouple and, thus, the comparison of the temperature measurements in the same position is not easy at all. Additionally, the uncertainty of the temperature measurement of the thermocouple has to be taken into account, as well as that of the IR camera; other issues to be considered when evaluating the IR camera measurement are the reflection of the shelf on the bottom of the vial and the dark shadow created where the vials get in contact. Therefore, in Figure 4.2 we compared the thermocouple measurement with that provided by the IR camera in 5 points where the tip of the thermocouple should be. It is possible to see that the thermocouple measurement comprises the range of values detected by the IR camera, and thus, the agreement is considered to be satisfactory. It can also be highlighted that the IR camera measure in position #3, where the thermocouple is supposed to be, is in excellent agreement with that of the thermocouple. Velardi and Barresi (2008)

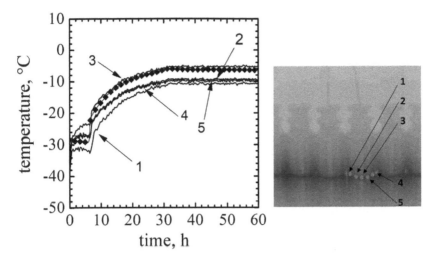

FIGURE 4.2 Comparison of the thermal measurements extracted from a thermocouple inserted inside a vial during the primary drying of a 5% b.w. sucrose solution, processed at −20°C and 20 Pa, and the temperature profiles obtained from 5 pixels at the bottom of the vial.

presented a detailed one-dimensional model describing the dynamics of the heat transfer inside the glass wall. The glass has a minor effect on the product dynamics, and the temperature difference between the external and the internal surface of the vial was generally in the range 0.1 to 0.7°C. Van Bockstal et al. (2018) proposed a simplified equation, Equation 4.1, that, once the heat flow from the environment has been characterized, could provide a direct correction of the temperature readouts:

$$T_p = T_{g,out} - \frac{\left(R+s_g\right)\ln\left(\dfrac{R+s_g}{R}\right)}{k_g}\dot{q}, \qquad (4.1)$$

where $T_{g,out}$ is the temperature on the outer side of the vial, T_p is the temperature of the glass in direct contact with the product, R is the vial radius, s_g is the thickness of the vial wall, k_g is the thermal conductivity of the glass, and $\dot{q}$ the heat flux from the environment to the vial. Using Equation 4.1 a temperature gradient of 0.88°C at the beginning of the primary drying and 0.47°C at the end was reported.

Equation 4.1, which is derived from a one-dimensional steady-state Fourier equation in the radial direction, neglects any kind of axial gradients; thus, its validity in the case of a batch of vials is questionable. The contact with the shelf will in fact induce some axial gradients that are not considered in Equation 4.1. Furthermore, in all those cases where the assumption of steady-state heat transfer might fail, such as the beginning and ending point of the process, the results obtained from Equation 4.1 must be carefully handled.

4.2.2 SUBLIMATION INTERFACE MONITORING—END
OF PRIMARY DRYING ESTIMATION

The published application of IR imaging technologies mainly focused on the monitoring of the primary drying stage. During this stage, an adequate process analytical technology (PAT) should constantly monitor the temperature of the product, thus assessing if the threshold for product denaturation, amorphous products collapse, or crystalline products melting is trespassed or not (Bellows and King 1972; Tsourouflis et al. 1976). An estimation of the residual amount of frozen solvent, to avoid unnecessary extension of the process, is also valuable. This will also protect the product quality avoiding the temperature to be increased when not all the ice has been sublimated.

Given the thermal picture of a vial, once the region corresponding to the product has been identified (see Figure 4.3A), the axial thermal profiles in many different positions could be extracted and used to track the thermal evolution of the product along the process. The maximum temperature can be monitored in real time, and in general, this will provide information about any abnormal thermal evolution. Even more important is the possibility to track the local minimum in these axial thermal profiles and its evolution over time; see Figure 4.3B.

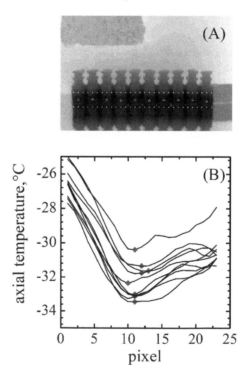

FIGURE 4.3 Details on the working principle of the algorithm for real-time monitoring of the position and temperature of the sublimation front. In Figure 4.3A each couple of white dots delimits the pixel where temperature profiles were extracted, while the white triangles report the actual position of the minima detected by the algorithm.

This minimum is strictly related to the position of the sublimation interface, thus providing both the information on the temperature at this point and its position along the height of the product, that is, the residual amount of ice. Figure 4.3B reports the temperature axial profiles obtained averaging three rows of pixels, extracted along the height of the product, for each one of the ten vials processed. Averaging multiple data acquisitions is a common strategy to filter the noise of measurement and enhance the quality of the readout. In Figure 4.4A the profiles of the sublimation front, H_i, for each one of the ten vials of a single batch (10R filled with 5 ml of a 5% b.w. sucrose solution with shelf temperature $-10°C$ and chamber pressure 20 Pa) are reported together with the one obtained from their average.

The actual position of the sublimation front z_i has been measured in pixels, but, to ease comprehension, the dimensionless value $H_i = \dfrac{z_i}{L}$ is reported, where L is the total height of the product. Figure 4.4B presents the values of the temperature at the sublimation front for all the vials monitored. These trends are in good agreement with those reported in literature, for example, the thickness of the frozen layer decreases very slowly at the beginning, when the pressure is reduced before the shelf temperature is increased, and then the rate of removal of the ice is almost constant until the sublimation front reaches the bottom of the vial (Velardi and Barresi 2008;

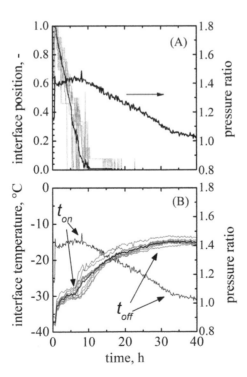

FIGURE 4.4 Ice front position (A) and temperature (B) for each one of the ten vials of a batch, compared with the resulting Pirani-Baratron ratio, and mean values for both variables (black solid lines). Each one of the 10R vials was filled with 5 ml of a 10% b.w. sucrose solution. Fluid temperature, $-20°C$; chamber pressure, 20 Pa.

Colucci and Fissore 2019). After the shelf temperature has been raised, T_i reaches a first asymptotical value, corresponding to an equilibrium between the heat supplied and the one removed from the ice sublimation. In some vials receiving much more heat because of their position in the batch, this asymptotical value is slightly higher than that in other vials. Then, at a certain moment, T_i starts increasing again because ice sublimation is (almost) completed and the heat provided by the shelf is being accumulated in the dried product (and no more completely removed through ice sublimation). This occurs after about 7 hours from the onset of the primary drying stage in those vials whose temperature is higher, and after about 9 hours in the rest of the batch. H_i presents a slope change after about 9 hours from the beginning of the primary drying stage, corresponding to the onset of the Pirani-Baratron ratio.

At the end of the primary drying we know there will be no humidity left inside the drying chamber. Thus, the measurement of the capacitive Baratron manometer and of the conductive Pirani gauge should be equal; their ratio will approach the unit. Seen from the point of view of the thermal evolution of the product when no ice is left inside the product, no further heat can be removed via sublimation; thus, the product will reach thermal equilibrium with the surrounding environment. These two points describe the rationale behind the definition of the offset of the primary drying and were used to obtain this value from the experimental measurement; see Figure 4.4B. The estimated values of the onset (t_{on}) and the offset (t_{off}) of the decrease of the pressure ratio curve are reported in the parity plot of Figure 4.5.

The noise of the measurement makes the estimation of the instant the thermal profile reaches a constant value very difficult and, for this reason, the offset appears to be systematically slightly underestimated by our algorithm.

In Lietta et al. (2019) the temperature read by the sensor at the bottom of the vial was proved to allow an estimation of the ending time of the primary drying stage within the same degree of uncertainty obtained using the Pirani-Baratron ratio. Van Bockstal et al. (2018) compared the temperature profile obtained by averaging all the pixels included in a box at the center of the product with the results obtained from the analysis of the NIR spectra. When the temperature started increasing, the spectra read from the NIR demonstrated the presence of residual ice inside the product.

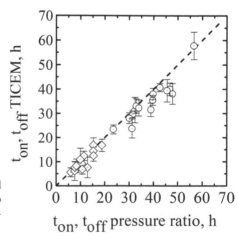

FIGURE 4.5 Parity plot for the onset and offset obtained from the Pirani-Baratron ratio and from the thermal profiles of the temperature at the sublimation front.

4.2.3 HEAT AND MASS TRANSFER COEFFICIENTS

The on-line estimation of K_v and R_p, respectively, the heat and mass transfer coefficients of the VFD process, has a dramatic importance. Many approaches have been presented for this task, including real-time soft sensors (Velardi et al. 2009, 2010; Bosca et al. 2015), but the most straightforward one is the one proposed by Fissore et al. (2017). Under the hypothesis that all the heat provided is used for ice sublimation and that heat is supplied only from the technical fluid, given a measurement of the temperature at the bottom of the vial, T_B, and the temperature of the fluid flowing inside the shelves, T_s, Equation 4.2 can estimate the heat, Q, provided to the product between t_0 and t_f :

$$Q = K_v A_v \int_{t_0}^{t_f} (T_s - T_b) \, dt, \tag{4.2}$$

where K_v is the heat transfer coefficient (see also Chapter 1.4) and A_v is the area of the vial available for the heat transfer.

Since we know the total amount of ice that sublimates, m_w, the total heat that must be supplied reads $Q = \Delta H_s \cdot m_w$, and it is possible to measure K_v. If we assume the thermal gradients inside the frozen layer to be negligible, it is possible to estimate the water partial pressure at the sublimation interface, $p_{w,i}$ (Goff and Gratch 1946). If the water partial pressure in the chamber, $p_{w,c}$, is known, together with the heat flux, estimated from the difference between the temperature of the fluid and that at the bottom of the product and K_v, we use Equation 4.3 to determine R_p:

$$R_p = \frac{\Delta H_s \left(p_{w,i} - p_{w,c} \right)}{K_v \left(T_s - T_B \right)}. \tag{4.3}$$

Once the profile of R_p over time has been calculated, we can fit the constants A and B that describe its dependence from the thickness of the dried product layer L_d:

$$R_p = \frac{A \cdot L_d}{1 + B \cdot L_d}. \tag{4.4}$$

The reliability of the measurement obtained was proven by applying this methodology to a large set of experiments, where different variables, namely temperature of the heating shelf ($-10°C$, $-20°C$, and $-30°C$), the pressure in the chamber (10 Pa and 20 Pa), the kind of sugar (sucrose and mannitol), the solid content in the solution (5% b.w. and 10% b.w.), and the vial geometry (4 R and 10 R) were investigated. In all tests, 10 vials (ISO 8362–1) were placed at 30 cm from the sensor described by Lietta et al. (2019), and the heat transfer coefficient, K_v, was measured.

A multiway ANOVA was performed to assess the statistical significance of the variables studied on the average K_v. Both the effects of temperature and pressure were statistically significant (p value < 0.05): K_v decreases with temperature, Figure 4.6A, and increases with pressure, Figure 4.6B. As expected, the amount of solid in the solution and the kind of sugar did not show a statistical effect (Figure 4.6C and Figure 4.6D, respectively). The kind (geometry) of vial was not

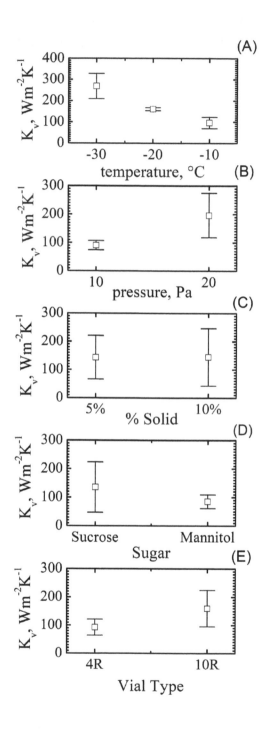

FIGURE 4.6 Heat transfer coefficient, K_v, between fluid and product as a function of temperature (A), pressure (B), percentage of solid in solution (C), sugar type (D) and vial type (E).

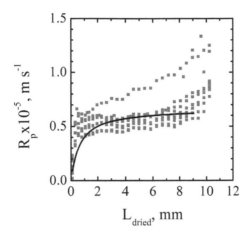

FIGURE 4.7 Experimental and fitted mass transfer coefficient R_p for all the ten vials of a single batch. Ten 10R vials filled with 5 ml of a 10% b.w. sucrose solution. Fluid temperature, −10°C; chamber pressure, 10 Pa.

statistically significant either in this case. These results agree with those presented in the theoretical study of Pisano et al. (2011a).

Figure 4.7 reports an example of experimental and interpolated profile of R_p for a batch obtained processing ten 10 R vials filled with 5 ml of a 5% b.w. sucrose solution with shelf temperature −10 °C and chamber pressure 10 Pa.

The experimental data were interpolated according to Equation 4.2, and the values of A and B obtained for this specific experimental condition are 2×10^8 s^{-1} and 1500 m^{-1}, respectively. The mass transfer coefficient, R_p, was found to be significantly dependent (p value < 0.05) on the kind of sugar and the amount of solid, while the effects of temperature, pressure, and vial geometry were negligible.

In the study by Van Bockstal et al. (2018), the temperature profile extracted from the thermal images was used to infer R_p, using Equation 4.5:

$$R_p = \frac{\Delta H_s A_p \left(p_{w,i} - p_{w,c} \right)}{P_{tot} M_w},$$ (4.5)

where ΔH_s is the molar latent heat of sublimation of the ice, A_p is the surface area available for sublimation, M_w is water molecular weight, and P_{tot} the total power supplied to the spin freezing product via radiation from the surroundings. The temperature of the product, required to estimate both $p_{w,i}$ and P_{tot}, was obtained from the infrared images.

4.3 MULTIVARIATE IMAGE ANALYSIS FOR STATISTICAL PROCESS MONITORING

One of the main advantages of an IR imaging system is that a full thermal characterization of the product, not a single measurement in a defined position, is provided. When reading the temperature of a single pixel, or a bunch of them, a lot of useful information will be discarded. Furthermore, one of the main sources of in-batch variability is the different drying conditions in different spatial positions in the chamber. Using a thermal image, we can monitor several vials at the same time and take into account their actual position inside the drying chamber. However, everything apart from the product is useless and should not be considered to avoid the introduction of noise into the signal.

As an example, Figure 4.8A is 320 × 256 pixels total, while the whole region corresponding to the product inside a single vial is approximately 400 pixels, thus the product inside all ten vials, the black boxes in Figure 4.8C, counts for roughly 5% of the total. In the algorithm presented in chapter 4.2.2 to monitor the position and temperature of the sublimation front, only 60 pixels (over the total 400 available) per vial were used: 85% of the data are discarded together with any possible information included. Indeed, the correct pretreatment of the images obtained has a tremendous effect on the amount and quality of the results that we can obtain. Image analysis, namely gray-scale image segmentation, image registration, optical effects correction, and so forth (Gonzalez et al. 2004), but also computer vision, that is, object detection and particle-tracking algorithms, is importance in this framework. Figure 4.8 shows an example of the image pretreatment performed on both thermal and RGB images during the online monitoring of a VFD process (Colucci et al. 2019a, 2019b). Figure 4.8 (A) is the original thermal image. Figure 4.8 (B) has been corrected for the optical aberration, and the position of the line of vials was identified. In Figure 4.8 (C) the region of each one of the ten vials corresponding to

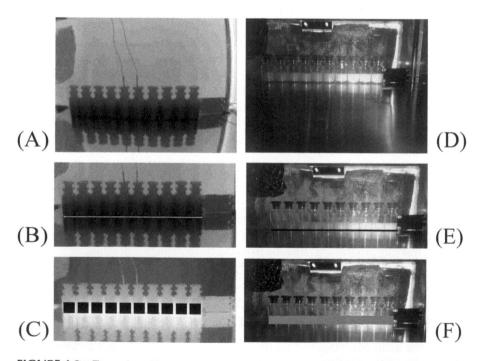

FIGURE 4.8   Examples of image pretreatment. (A), original RGB image; (B), object detection; (C), resulting segmentation of the product inside the vials; (D), original IR image; (E), optical distortion correction and object detection; (F), resulting segmentation.

the product has been segmented. Similar results are reported for the RGB images in Figure 4.8 (D), the original one; Figure 4.8 (E), object identification and cropping; and Figure 4.8 (F), resulting segmentation.

Another major issue is the high amount and low quality of the data extracted, which will require the right analytical tools to extract all the information available. A first problem is the noise of the measurement, which might be very high in this kind of systems owing to all the limitations that must be addressed to obtain a good measurement. Usually, the use of high-resolution systems or the acquisition of the average of multiple images, obtained at the maximum rate allowed by the system, is enough to increase the signal to noise ratio. Filters should be applied carefully in order to not filter the valuable information. The correlation of the data structure obtained is another major drawback. Since the pixels are next to each other, the single pixel temperature cannot be considered independent from that of his neighbours, they are said to be spatially correlated. Additionally, the measurement of the same pixel at different times is strongly related to the values obtained at the previous time steps. This characteristic of an experimental measurement is called *auto-correlation* and is typical of batch systems, such as VFD. Latent variables–based multivariate statistical techniques can easily deal with these kinds of problems. Kourti (2006) discussed the primary role of multivariate statistical techniques in the development of PAT for the pharmaceutical industry. The application of multivariate statistical techniques to the extraction of information from images falls inside the framework of the multivariate image analysis (MIA) (Prats-Montalbán et al. 2011; Duchesne et al. 2012). In MIA two main approaches can be distinguished: a "pixel" level and a "global image" level. In the former the intensity of a single pixel is assumed as an observation of the whole image. Notice that in a thermal image, intensity is equivalent to temperature. The latter focuses not on the characteristics of a single pixel but on those features extracted from a group of pixels of the image. In the study by Colucci et al. (2019a) the temperature distribution extracted from each of the regions belonging to the product inside the vials, see the segmentation in Figure 4.8 (C), was characterized by extracting the mean, standard deviation, skewness, and kurtosis for each one of the images acquired for a single batch. A global image approach was preferred to ease the handling of such an amount of data and filter some of the noise of the measurement without losing information about the thermal evolution inside the product.

The idea underlying the development of a latent variable multivariate monitoring system is that the information included into a data set of normal operating conditions (NOCs) could be compressed into a lower number of fully independent variables. These new "latent variables" describe the few driving forces that are governing the process. In the reduced-dimension space described by this new set of variables, we can build control charts able to monitor the evolution of the process and identify any fault eventually occurring. Both principal component analysis (PCA) and partial least squares (PLS) have been widely studied and applied for this purpose (Nomikos and MacGregor 1994, 1995a; Kourti 2005). Latent variable multivariate monitoring systems were proposed by Nomikos and MacGregor (1995b) and adapted to the peculiarities of a batch process by Kourti (2003). In recent years, many upgrades of the original ideas were proposed and tested (Ramaker et al. 2002) and many methodologies for a systematic comparison of the different approaches have been presented (Van Sprang et al. 2002; Aguado et al. 2007; Rato et al. 2016, 2018). Duchesne et al. (2012) published a full review of the industrial applications of MIA, including latent variable multivariate system for real-time monitoring, optimization, and control.

The correct pretreatment of the data, before the latent variable model is extracted, has a relevant effect on the quality of the obtained results. The measurement of a certain number of variables (J), for a certain number of observations (I) and, for a batch process, along K instants of time, generates a three-dimensional data structure. Both PCA and PLS, in their original versions, can only deal with two-dimensional data structures; thus, the original data must be unfolded. Batch-wise unfolding (BWU), that is, the features extracted for a single batch are put beside each other in order of time acquisition to obtain a matrix $\mathbf{X}(I \times JK)$ (Kourti 2003), was shown to be the appropriate approach for modeling the time-varying batch process dynamics (Camacho et al. 2008, 2009). Usually all the variables are trajectory centered (i.e., the average trajectory is removed) and scaled to unit variance to give all the time instants the same "a priori" importance into the model, no matter their original differences in variability.

Given a general two-dimensional matrix, $\mathbf{X}(I \times JK)$, a PCA model performs a bilinear decomposition of the data structure, in the form:

$$\mathbf{X} = \mathbf{T} \cdot \mathbf{P}^{\mathrm{T}} + \mathbf{E}, \tag{4.6}$$

where $\mathbf{T}$ is the $I \times A$ score matrix, $\mathbf{P}$ is the $JK \times A$ loading matrix, and $\mathbf{E}$ is the residual matrix, having the same dimension as the original matrix $\mathbf{X}$, and A being the number of latent variables extracted.

Similarly, a PLS model with A latent variables has the following structure:

$$\mathbf{T} = \mathbf{X}\,\mathbf{W}^*, \tag{4.7}$$

$$\mathbf{X} = \mathbf{T} \cdot \mathbf{P}^{\mathrm{T}} + \mathbf{E}, \tag{4.8}$$

$$\mathbf{Y} = \mathbf{T} \cdot \mathbf{C}^{\mathrm{T}} + \mathbf{F}, \tag{4.9}$$

where $\mathbf{T}$ is the $I \times A$ score matrix of $\mathbf{X}$, $\mathbf{P}$ is the $JK \times A$ loading matrix of $\mathbf{X}$, $\mathbf{W}^*$ is the $JK \times A$ weighting matrix of $\mathbf{X}$, and $\mathbf{C}$ is the corresponding $J^* \times A$ weighting matrix of $\mathbf{Y}$. $\mathbf{E}$ and $\mathbf{F}$ are the matrices of the residuals, having the same dimension of the matrices $\mathbf{X}$ and $\mathbf{Y}$, respectively. The matrix $\mathbf{Y}$ is also usually mean centered and scaled to unitary variance. Using the PLS it is possible to obtain a real-time prediction of the final quality attributes of the product, Equation 4.10, and following the procedure presented by Nomikos and MacGregor (1995a), we can also build the control charts of these predicted values:

$$\mathbf{Y}_{\mathrm{pred}} = \mathbf{X} \cdot \mathbf{W}^* \cdot \mathbf{C}^{\mathrm{T}}. \tag{4.10}$$

In both PCA- and PLS-based multivariate statistical process control (MSPC) the reference model is created using only the data obtained from the NOC data set. Then, the vector of the new observation (i.e., the new batch) is projected into the subspace defined by the latent variables. Here the information is finally compressed into two independent indexes: Hotelling T^2 (T^2) and the instantaneous square prediction error at time k ($SPEI_k$), defined by Equations 4.11 and 4.12:

$$T^2 = \sum_{a=1}^{A} \frac{t_a^2}{\lambda_a},$$

(4.11)

$$SPEI_k = \sum_{c=1+(k-1)J}^{Jk} e_c^2,$$

(4.12)

where t_a is the a-th score, λ_a is the eigenvalue of the a-th principal component, e_c is the error obtained after predicting the latest online measurements at time k for a certain observation and k the actual time instant at which the measurement is performed.

Using the distributions of these indexes obtained from the training data set, it is possible to calculate the upper control limits (UCLs) of the control charts (Kourti and MacGregor 1996). When an observation trespasses the UCL, the contribution plots can be used to figure out which of the original variables are responsible for this deviation from the evolution of the NOC data. More details about the algorithms and how these statistical techniques can be used to perform online monitoring and fault detection and their pros and cons can be found in the article by Nomikos (1995).

It is mandatory to build the model for the algorithm using only observations that describe the desired evolution of the process, namely the normal operating conditions (NOCs). When the "usual" correlation structure has been modelled, any difference from this standard evolution will appear as an abnormal behavior in the control charts. The most important step of the development of a latent variable multivariate monitoring system, though, is the validation of the model, that is, the assessment of the performances and the quality of the classification obtained. A new set of observations, both NOC and faults, are projected onto the previously obtained model trying to adjust the control limits, the number of latent variables extracted, and the kind of data pretreatment applied, in order to minimize the errors in the classification of the observation, namely false positives and false negatives. A false positive is a NOC batch classified as a fault, while a false negative is a faulty batch not trespassing the control limits of the control charts and, therefore, considered a NOC one.

In the article by Colucci et al. (2019a) an application of these multivariate statistical techniques to the real-time monitoring of a VFD process was presented, and different approaches were tested to obtain the best performances in terms of fault detection. The two main points addressed in this paper regarded the inclusion of the position of the single vial of the batch among the monitored variables and the best algorithm to be used for the prediction of the missing information. One of the main sources of the in-batch variability was due to the spatial position of the different vials in the chamber. A thermal image provides information on more vials placed in different positions; therefore, a question arises: is it better to deal with a "vial-wise" system, that is, each vial is considered as an observation on its own, thus completely independent from the others and from the environment, or shall we consider the whole batch as an observation? The results presented clearly show that the latter approach makes the fault detection more robust and reliable owing to the inclusion of the spatial information. This approach has two minor drawbacks. The detection of the faults occurring to a single vial has to be entrusted to the contribution plots (bar diagrams reporting the contribution of each variable to the total value of the *SPEI* and T^2) and the dimension of the matrix has N times more columns and N times less observations than the matrix obtained with the "vial-wise" approach. N is the number of vials monitored in the batch.

The second problem is typical of every real-time application of these techniques to time-varying systems: when the k-th image is acquired ($k < K$) the future part of the trajectory of each variable j is missing and must be completed in some way (Nomikos and MacGregor 1995a). Arteaga and Ferrer (2002, 2005) showed that the most statistically efficient methods to this purpose are those that estimate the scores for the new incomplete observation as the prediction from a regression model: the so-called regression-based methods. In Arteaga and Ferrer (2002, 2005) the trimmed score regression (TSR) method and the known data regression (KDR) were tested and compared following the approach proposed by Garcia-Muñoz et al. (2004) that evaluates the accuracy of the scores estimation, the quality of the prediction of the missing information, and the detection ability. The TSR outperforms the KDR in terms of both calculation time (especially when many variables are accounted) and quality of the estimation of the unknown observations, providing a more responsive SPEI control chart and a better classification performance. The KDR provides better quality scores, but this improvement is not enough to compensate for the intrinsically lower reliability of the T^2 control chart (Camacho et al. 2008). Figure 4.9 shows an example of the control charts obtained for one NOC (in gray) and one faulty observation (in dark gray).

In the paper by Colucci et al. (2019b) other variables providing information complementary to that of the features extracted from the thermal images (measurements obtained directly from the PLC of the equipment, and a textural index extracted from the analysis of the RGB images) were included. When different blocks of variables are added to the original data matrix, the possible strategies for data pretreatment increase, since relative weight of each block might be adjusted to improve the possible results. When all the variables are unit scaled, all of them contribute to the model in the same way, so it would be possible that some relevant variables of a small group lose weight in the model in comparison with other nonrelevant variables of a larger group. To avoid these inconveniences, the data can be block scaled (Westerhuis et al. 1998), so that the same a priori weight is given to different blocks of variables, no matter their size. In the cited paper, the base case (B), where only the features extracted from the thermal images were considered, was compared with the case where three different kinds of block scaling were applied. In the first case, referred to as (N), no block scaling was applied, that is, the two blocks contribute to the final model according to the number of variables in each one of the original

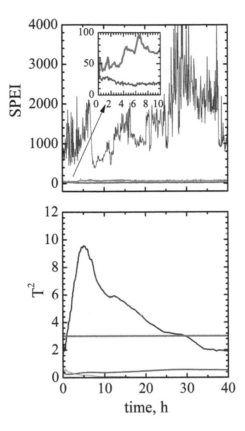

FIGURE 4.9 Instantaneous squared prediction error (SPEI) and Hotelling T^2 (T^2) control charts. Red line, upper control limits; dark gray line, fault; gray, NOC observation; light gray, fault detected by the SPEI, false negative according to the T^2 control chart.

groups. A second approach tested was to give both the blocks a unitary variance (U), while in the last approach (I) only the thermal information was scaled to a unitary variance, emphasizing the contribution of the additional variables.

The performances obtained with both PCA and PLS were studied and compared. Given a certain number of latent variables, and once the kind of block scaling strategy to be applied was defined, the performances of the algorithms were evaluated according to Equation 4.13:

$$f_A = (1 - \alpha) \cdot (1 - \beta), \tag{4.13}$$

where α is the overall type I error rate (i.e., false-positive rate) and β is the overall type II error rate (i.e., false-negative rate) (Prats-Montalbán and Ferrer 2014). The lower α and β, the more f_A approaches the unit.

Figure 4.10 summarizes these results.

Both algorithms provide good performance, which can be improved as more latent variables are extracted and with the inclusion of additional variables and the kind of block scaling used. The perfect classification of all the batches was achieved

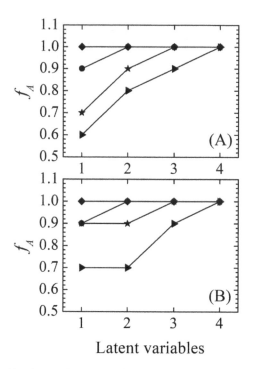

Latent variables

FIGURE 4.10 Classification performance of the PCA-based (A) and PLS-based (B) MSPC algorithms with and without additional variables and for different block-scaling approaches. Triangles, no additional variable in the X matrix; stars, thermal and additional variables without any block scaling (N); circles, thermal and additional variables scaled at unitary variance (U); rhombus, inverse scaling of the two blocks of data (I).

with models built extracting only one or two latent variables when a (U) or (I) block scale was applied.

4.4 CONCLUSIONS

Infrared imaging has proven in the recent years to be the raising star of the vacuum freeze-drying process monitoring. A full thermal characterization of the product, and its evolution over time, can be obtained without interfering with the dynamics of the product. The possibilities deriving from the application of some basic image analysis to extract the information embedded in the huge amount of data usually provided are outstanding and will definitely call the attention of both pharmaceutical companies and freeze-drying practitioners. In this chapter, the main applications of this PAT proposed in the literature have been reviewed and discussed. They mainly focus on the real-time monitoring of the process, especially primary drying. Future work will have to deepen the possible applications of this technology to the study of the other phases of the VFD process, in particular, given its dramatic importance, the study and monitoring of the freezing step. Another relevant possibility is the implementation of the proposed technology into control algorithms, either the simplest

PID controller or most sophisticated MPC algorithms, might take advantage of the information about both the coefficients required to simulate the process and the measurements of the position and temperature of the freezing or sublimating interface. In this framework, the multivariate statistic techniques previously discussed might be soundly beneficial. In fact, even the simplest model of the VFD process requires the solution of a nonlinear set of differential equations, while PCA and PLS can model this nonlinearity in the form of very simple algebraic system of equations. From the technical point of view the implementation of a sensor based on an infrared camera, especially in the case of a batch process and in production plants, still presents some challenges, such as sterilization, miniaturization, and algorithms optimization, that must be addressed. It is our opinion that the possible benefits that could be obtained will drive this part of the research, which will have to include expertise from many different fields.

REFERENCES

Aguado, D., Ferrer, A., Ferrer, J., and A. Seco. 2007. Multivariate SPC of a sequencing batch reactor for wastewater treatment. *Chem. Intell. Lab. Syst.* 85:82–93.

Albanese, D., Cinquanta, L., Cuccurullo, G., and M. Di Matteo 2013. Effects of microwave and hot-air drying methods on color, β-carotene and radical scavenging activity of apricots. *Int. J. Food Sci. Technol.* 48:1327–1333.

Arteaga, F., and A. Ferrer. 2002. Dealing with missing data in MSPC: Several methods, different interpretations, some examples. *J. Chemom.* 16:408–418.

Arteaga, F., and A. Ferrer. 2005. Framework for regression-based missing data imputation methods in on-line MSPC. *J. Chemom.* 19:439–447.

Bagavathiappan, S., Lahiri, B. B., Saravanan, T., Philip, J., and T. Jayakumar 2013. Infrared thermography for condition monitoring – A review. *Infrar. Phys. Tech.* 60:35–55.

Bellows, R. J., and C. J. King. 1972. Freeze-drying of aqueous solutions: Maximum allowable temperature. *Cryobiology* 9:559–961.

Bosca, S., and D. Fissore. 2011. Design and validation of an innovative soft-sensor for pharmaceuticals freeze-drying monitoring. *Chem. Eng. Sci.* 66:5127–5136.

Bosca, S., Barresi, A. A., and D. Fissore. 2014. Use of soft-sensors to monitor a pharmaceuticals freeze-drying process in vials. *Pharm. Dev. Tech.* 19:148–159.

Bosca, S., Barresi, A. A., and D. Fissore. 2015. Design of a robust soft-sensor to monitor in-line a freeze-drying process. *Drying Technol.* 33:1039–1050.

Bosca, S., Corbellini, S., Barresi, A. A., and D. Fissore. 2013. Freeze-drying monitoring using a new process analiytical technology: Toward a "zero defect" process. *Drying Technol.* 31:1744–1755.

Böttger F, and M. Häuptle. 2009. Large-scale lyophilization device. U.S. Patent Application 20110099836 filed June 16, 2009.

Brulls, M., Folestad, S., Sparén, A., and A. Rasmuson. 2003. In-situ near-infrared spectroscopy monitoring of the lyophilization process. *Pharm. Res.* 20:494–499.

Camacho, J., Picó, J., and A. Ferrer. 2008. Multi-phase analysis framework for handling batch process data. *J. Chemom.* 22:632–643.

Camacho, J., Picó, J., and A. Ferrer. 2009. The best approaches in the on-line monitoring of batch processes: Does the modelling structure matters? *Analyt. Chim. Acta* 642:59–68.

Colucci, D., and D. Fissore. 2019. Thermal images based on-line monitoring of the freeze-drying process. Paper presented at the third Nordic Baltic Drying Conference (NBDC 2019), Saint Petersburg, Russia, June 12–14.

Colucci, D., Prats-Montalbán, J. M., Fissore, D., and A. Ferrer. 2019a. Application of multivariate image analysis for on-line monitoring of a freeze-drying process for pharmaceutical products in vials. *Chemom. Intell. Lab. Syst.*187:19–27.

Colucci, D., Prats-Montalbán, J. M., Ferrer, A., and D. Fissore. 2019b. On-line product quality and process failure monitoring in freeze-drying of pharmaceutical products. *Drying Technol.* In press (DOI: 10.1080/07373937.2019.1614949).

Corbellini, S., Parvis, M., and A. Vallan. 2010. In-process temperature mapping system for industrial freeze-dryers. *IEEE Trans. Inst. Meas.* 59:1134–1140.

Cuibus, L. 2013. Applications of infrared thermography in the food industry. PhD diss., Università di Bologna.

De Beer, T., R., Vercruysse, P., Burggraeve, A., et al. 2009. In-line and real-time process monitoring of a freeze-drying process using Raman and NIR spectroscopy as complementary process analytical technology (PAT) tools. *J. Pharm. Sci.* 98:3430–3446.

Demichela, M., Barresi, A. A., Baldissone, G. 2018. The effect of human error on the temperature monitoring and control of freeze drying processes by means of thermocouples. *Front. Chem.* 6:11 pp.

Duchesne, C., Liu, J. J., and J. F. MacGregor. 2012. Multivariate image analysis in the process industries: A review. *Chemom. Intell. Lab. Syst.* 117:116–128.

Emteborg, H., Zeleny, R., Charoud-Got, J., et al. 2014. Infrared thermography for monitoring of freeze-drying processes: Instrumental developments and preliminary results. *J. Pharm. Sci.* 103:2088–2097.

Fissore, D., Pisano, R., and A. A. Barresi. 2017. On the use of temperature measurement to monitor a freeze-drying process for pharmaceuticals. I2MTC – 2017 IEEE International Instrumentation and Measurement Technology Conference – 2017 Proceedings Papers, May 22–25, Torino, Italy, pp. 1276–1281.

Fissore, D., Pisano, R., and A. A. Barresi. 2018. Process analytical technology for monitoring pharmaceutical freeze-drying – A comprehensive review. *Drying Technol.* 36:1839–1865.

Fito, P. J., Ortolá, M. D., De los Reyes, R., Fito, P., and E. De los Reyes. 2004. Control of citrus surface drying by image analysis of infrared thermography. *J. Food Eng.* 61:287–290.

Garcia-Muñoz, S., Kourti, T., and J. F. MacGregor. 2004. Model predictive monitoring for batch processes. *Ind. Eng. Chem. Res.* 43:5929–5941.

Goff, J. A., and G. Gratch. 1946. Low-pressure properties of water from -160 to 212 °F. *Trans. Amer. Soc. Heat. Vent. Eng.* 52:95–122.

Gonzalez, R. C., Woods, R. E., and S. L. Eddins. 2004 *Digital image processing using Matlab.* Upper Saddle River: Pearson education, Inc.

Grassini, S., Parvis, M., and A. A. Barresi. 2013. Inert thermocouple with nanometric thickness for lyophilization monitoring. *IEEE Trans. Instrum. Meas.* 62:1276–1283.

International Organization for Standardization. ISO 18434–1. Condition monitoring and diagnostics of machines – Thermography – Part 1: General procedures. Geneva: ISO; 2008. www.iso.org/obp/ui/#iso:std:iso:18434:-1:ed-1:v1:en (accessed March 15, 2019).

Kourti, T. 2003. Multivariate dynamic data modeling for analysis and statistical process control of batch processes, start-ups and grade transitions. *J. Chemom.* 17:93–109.

Kourti, T. 2005. Application of latent variable methods to process control and multivariate statistical process control in industry. *Int. J. Adapt. Contr. Signal Proc.* 19:213–246.

Kourti, T. 2006. Process analytical technology beyond real-time analyzers: The role of multivariate analysis. *Crit. Rev. Analyt. Chem.* 36:257–278.

Kourti, T., and J. F. MacGregor. 1996. Multivariate SPC methods for process and product monitoring. *J. Qual. Tech.* 28:409–428.

Lahiri, B. B., Bagavathiappan, S., Jayakumar, T., and J. Philip. 2012. Medical applications of Infrared thermography: A review. *Infrar. Phys. Tech.* 55:221–235.

Lietta, E., Colucci, D., Distefano, G., and D. Fissore. 2019. On the use of IR thermography for monitoring a vial freeze-drying process. *J. Pharm. Sci.* 108:391–398.

Meola, C. 2012. Origin and theory of infrared thermography. In *Infrared thermography: Recent advances and future trends*, ed. C. Meola, 3–28. Soest: Bentham Science Publishers.

Nomikos, P. 1995. Statistical process control of batch processes. PhD diss., McMaster University, Hamilton.

Nomikos, P., and J. F. MacGregor. 1994. Monitoring batch processes using multiway principal component analysis. *AIChE J.* 40:1361–1375.

Nomikos, P., and J. F. MacGregor. 1995a. Multi-way partial least squares in monitoring of batch processes. *Chem. Intell. Lab. Syst.* 30:97–108.

Nomikos, P., and J. F. MacGregor. 1995b. Multivariate SPC charts for monitoring batch processes. *Technom.* 37:41–59.

Oetjen, G. W., and P. Haseley. 2004. Freeze-drying. Weinheim: Wiley-VHC.

Pillans, L. A. 2008. Performance evaluation of an uncooled infrared array camera. PhD diss., University of London, London.

Pisano, R., Fissore, D., and A. A. Barresi. 2011a. Heat transfer in freeze-drying apparatus. In *Developments in Heat Transfer*, ed. M. A. dos Santos Bernardes, 91–114. London: InTech.

Prats-Montalbán, J. M., and A. Ferrer. 2014. Statistical process control based on multivariate image analysis: A new proposal for monitoring and defect detection. *Comp. Chem. Eng.* 71:501–511.

Prats-Montalbán, J. M., De Juan, A., and A. Ferrer. 2011. Multivariate image analysis: A review with applications. *Chem. Intell. Lab. Syst.* 107:1–23.

Ramaker, H. J., Van Sprang, E. N. M., Gurden, S. P., Westerhuis, J. A., and A. K. Smilde. 2002. Improved monitoring of batch processes by incorporating external information. *J. Proc. Contr.* 12:569–576.

Rato, T. J., Rendall, R., Gomes, V., et al. 2016. A systematic methodology for comparing batch process monitoring methods: Part I – Assessing detection strength. *Ind. Eng. Chem. Res.* 55:5342–5358.

Rato, T. J., Rendall, R., Gomes, V., Saraiva, P. M., and M. S. Reis. 2018. Systematic methodology for comparing batch process monitoring methods: Part II – Assessing detection speed. *Ind. Eng. Chem. Res* 57:5338–5350.

Samanian, N., and M. Mohebbi. 2016. Thermography, a new approach in food science studies: A review. *MOJ Food Proc. Tech.* 2:110–119.

Schneid, S., and H. Gieseler. 2008. Evaluation of a new wireless temperature remote interrogation system (TEMPRIS) to measure product temperature during freeze drying. *AAPS PharmSciTech* 9:729–739.

Tsourouflis, S., Flink, J. M., and M. Karel. 1976. Loss of structure in freeze-dried carbohydrates solutions: Effect of temperature, moisture content and composition. *J. Sci. Food Agric.* 27:509–519.

Van Bockstal, P. J., Corver, J., De Meyer, L., Vervaet, C., and T. De Beer. 2018. Thermal imaging as a noncontact inline process analytical tool for product temperature monitoring during continuous freeze-drying of unit doses. *Analyt. Chem.* 90:13591–13599.

Van Sprang, E. N. M, Ramaker, H. J., Westerhuis, J. A., Gurden, S. P., and A. K. Smilde. 2002. Critical evaluation of approaches for on-line batch process monitoring. *Chem. Eng. Sci.* 57:3979–3991.

Velardi, S. A., and A. A. Barresi. 2008. Development of simplified models for the freeze-drying process and investigation of the optimal operating conditions. *Chem. Eng. Res. Des.* 86:9–22.

Velardi, S. A., Hammouri, H., and A. A. Barresi. 2009. In line monitoring of the primary drying phase of the freeze-drying process in vial by means of a Kalman filter based observer. *Chem. Eng. Res. Des.* 87:1409–1419.

Velardi, S. A., Hammouri, H., and A. A. Barresi. 2010. Development of a high gain observer for in-line monitoring of sublimation in vial freeze-drying. *Drying Technol.* 28:256–268.

Willemer, H. 1991. Measurement of temperature, ice evaporation rates and residual moisture contents in freeze-drying. *Dev. Biol. Standard.* 74:123–136.

Westerhuis, J. A., Kourti, T., and J. F. MacGregor. 1998. Analysis of multiblock and hierarchical PCA and PLS models. *J. Chemom.* 12:301–321.

Zane, A. A. 2010. Using thermal imaging for control of a manufacturing process. U. S. Patent Application US20120057018A1 filed May 13, 2010.

5 Through-Vial Impedance Spectroscopy (TVIS)
A New Method for Determining the Ice Nucleation Temperature and the Solidification End Point

Geoff Smith and Yowwares Jeeraruangrattana

CONTENTS

5.1 AN INTRODUCTION TO THROUGH-VIAL IMPEDANCE SPECTROSCOPY

Through-vial impedance spectroscopy (TVIS), as the names suggests, measures the electrical impedance spectrum of a solution through the walls of the container in which the solution is freeze-dried. The location of the electrodes on the outside of the container means that the technology is not product invasive, and therefore the impact on nucleation, ice growth, heat transfer, and dry rates is minimized. With the current design of the TVIS impedance analyzer, a single pair or multiple pairs of electrodes are physically attached to the outer wall of the container (a glass vial or ampoule) (Figure 5.1).

The low thermal mass and streamlining of the external electrodes, which are connected to fine 'flexible' coaxial cables (not shown) means that the technology is also nonperturbing to the packing of the vials in the dryer (thereby maintaining the inherent heat transfer characteristics to the vials).

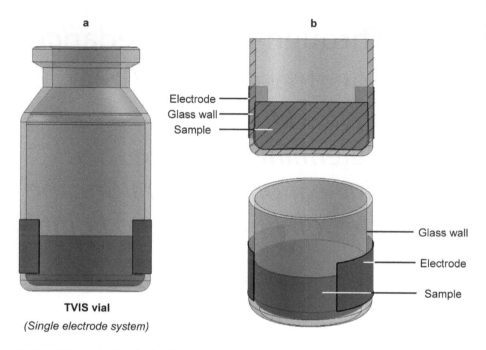

FIGURE 5.1   (a), Sketch of a TVIS vial with a pair of 19 × 10 mm electrodes attached to the external wall of the vial and positioned 3 mm from the base. (b), Three-dimensional sketch of the segment of glass and sample, which is measured in the TVIS impedance spectrum.

From an electrical impedance perspective, the TVIS vial can be regarded as a composite object, comprising two arc sections from the glass wall of a vial, in series, with a cylinder of ice (Figure 5.1b). The impedance of each component is represented by a complex capacitance ($C^* = C' - iC''$). The complex capacitance of the glass wall is characterized by (1) a frequency-dependent component (a low-frequency dispersion) that can be modelled over a limited range of frequencies by a constant phase element, and (2) a frequency-independent component modelled by a simple capacitor. The complex capacitance of the ice cylinder is characterized by (1) a dielectric relaxation that can be modelled by a Cole-Cole dispersion and (2) a frequency-independent component modelled by a simple capacitor. For both components, the frequency-independent capacitances model the instantaneous electronic and atomic polarizations of the glass wall and the ice, respectively.

Impedance spectra are generated throughout the drying cycle and displayed in terms of the two components of the equivalent complex capacitance spectra, that is, the real capacitance (dielectric storage) (Figure 5.2a) and imaginary capacitance (dielectric loss) (Figure 5.2b).

The example spectrum given here is for a TVIS vial containing ice at −20°C. The complex capacitance spectrum has a frequency response that is due to the low-frequency dispersion of the glass wall impedance at low frequencies (<100 Hz or log frequency = 2, region I in Figure 5.2). Then, as the impedance of the low-frequency dispersion decreases, and as the frequency of the excitation signal increases, the

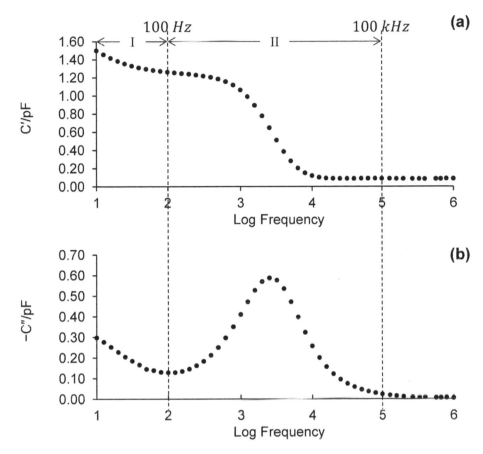

FIGURE 5.2 Real (a) and imaginary (b) components of the complex capacitance spectrum of a type I tubular glass vial with a pair of 10 × 19 mm electrodes attached to the external surface of vial and containing 3.5 g of double-distilled water that has been frozen and then reheated to −20°C.

impedance gives way to a pronounced dielectric relaxation of ice that dominates the majority of the frequency range of the TVIS complex capacitance spectrum. For the current design of analyser, this experimental frequency window extends from 10 Hz to 1 MHz. The dielectric relaxation of ice (whether measured through the glass vial or in the more conventional approach of using planar metal electrodes) spans three log decades in frequency (100 Hz to 100 kHz, region II in Figure 5.2) and is manifest as a step change in the real-part spectrum and a peak in the imaginary part spectrum. However, it is important to realise that, while the frequency dependence of the glass wall impedance dominates the visible characteristics of the low frequency part of the spectrum, the entire spectrum is also impacted by the series impedance of the glass wall and its interface with the contents of the vial. It follows that the real-part capacitance, in the limit of either low or high frequency relative to the relaxation frequency of ice, is a function of both the glass-wall/interfacial impedance and either the static capacitance of ice (in the case of the low frequency end of the spectrum) or the instantaneous capacitance of ice (in the case of the high frequency end of the spectrum).

One final point to make is that although the relaxation peak observed for frozen water is due to the dielectric relaxation of ice, additional process may also occur for frozen solutions. There are two potential mechanisms that underpin the physical origin of these additional processes: the first is the dynamic glass transition (the so called α-relaxation); the second is the Maxwell-Wagner (MW) or interfacial polarization, either at the glass-wall/solution interface or the interstitial spaces between the ice crystals. These additional processes may appear separated from and usually at lower frequencies than the main relaxation of ice or merged with the ice relaxation to form one larger peak, whereby this larger peak presents as either a peak with a low frequency shoulder (in the case of a partially merged peak) or as a single Debye-like peak (in the case of a fully merged peak). The proximity of these additional processes to the ice relaxation peak depends on the composition of the unfrozen fraction and the temperature of the frozen solution. In one published example, our observations appear to suggest that the temperature dependence of a fully merged process (e.g., that for a 10% w/v maltodextrin solution) tends to be associated with that of the unfrozen fraction rather than the ice fraction and so may be used to determine the glass transition temperature (Smith et al. 2013).

With water, and also with some solutions, it is often sufficient to follow the characteristic frequency and amplitude of the dielectric loss peaks in order to determine

1. the suppression of eutectics by glass-forming solutes (Arshad et al. 2014)
2. the changes in ice structures that occur on annealing from which the annealing hold time may be optimized (Smith et al. 2014)
3. the measurement of the average temperature of ice at a loci within the frozen volume (Smith et al. 2017)
4. the drying rate of ice and the temperatures of ice at the sublimation interface and at the interface with the base of the vial, from which the heat transfer coefficient may be derived (Smith et al. 2018)

In other cases (to be reported in future publications) it is necessary to use a dielectric relaxation model to fit the spectra and to separate out the contributions from the ice and the unfrozen factions. These features of impedance spectroscopy set it apart from other applications of impedance analysis, whereby a single frequency (often using 1 kHz) is used to investigate the phase changes of the solution to be freeze-dried, in an approach that is akin to that used in thermal analysis whereby the pre-frozen solution is reheated in order to observe its phase behaviour.

5.2   TVIS APPLICATIONS—AN OVERVIEW

The idea that it might be possible to measure the freeze-dried contents of a glass vial, with electrodes positioned on the outside of the vial, was first evaluated by by Suherman (2001) and Suherman et al. (2002). The materials under test were primarily hydrated proteins, for example, ovalbumin and lysozyme, and the 2–3 dielectric relaxations observed were considered to be due to either the Maxwell-Wagner polarization of the glass wall or the relaxations of localized proton percolation through the hydration shell of the protein. The work was influenced by Careri et al. (1985) who studied freeze-dried lysozyme in a glass dish with parallel plate electrodes. Later, stimulated by a collaboration between De Montfort University and GEA Pharma Systems, work started on the evaluation of the use of external electrodes to monitor the freeze-drying process (Innovate UK Collaborative R&D project grant, LyoDEA 100527). That project led to a bespoke impedance analyzer that was specifically designed for the purpose of measuring the dielectric properties of solutions undergoing a freeze-drying process from a glass vial (Smith and Polygalov 2019).

Applications for TVIS in the development of a freeze-drying process can be divided into qualitative studies and quantitative studies.

Qualitative studies largely involve the study of the onset or end point of some stage of the process, for example, (1) the ice nucleation event; (2) the completion of ice formation and/or the formation of eutectics during the solidification/freezing phase; (3) the completion of any recrystallization (Ostwald ripening) during an annealing stage; (4) the observation of collapse; and (5) the end point of primary drying.

Quantitative studies (which require some form of calibration) include: (1) the prediction of ice temperatures during primary drying; (2) the determination of the glass transition temperatures; (3) the determination of the sublimation rate in primary drying, from which critical process parameters may be derived, including the vial heat transfer coefficient and the product dry layer resistance. Other applications might be conceived for the secondary drying stage, but these will be the subject of future studies.

Figure 5.3 summarizes the basic responses of the TVIS dielectric loss spectrum of a water-filled vial to the processes of (a) cooling the liquid phase, (b) phase transition (liquid water to ice), (c) heating of the solid phase, and (d) primary drying.

Any changes in the dielectric loss peak due to the relaxation of ice are mirrored in the step in the dielectric storage (as was demonstrated in Figure 5.2), but given that it is easier to see the relative changes in the frequency and dielectric strength of a relaxation process in the form of a peak, then these first explorations of the potential applications for TVIS in Figure 5.3 show only the dielectric loss peak.

In the liquid state the dielectric loss peak appears at first glance to be similar to the dielectric relaxation of ice. However, this relaxation is due to an entirely different relaxation process known as a Maxwell-Wagner polarization or interfacial polarization, which occurs at the extremes of the physical geometry of the sample. In other words, there is an accumulation of charge (polarization) at the interface between the glass wall and the water (where the charges originate from mobile ionic species in the water). As stated earlier, the impedance of the glass wall is impacted by the interfacial layer that forms between the glass and the solvent, and so it follows that a study of this interfacial impedance provides one possible assessment of the nature of the interaction between different liquid and the glass.

The temperature dependence of the characteristic relaxation frequency (F_{PEAK}) of the Maxwell-Wagner relaxation (Figure 5.3a) is largely due to the temperature

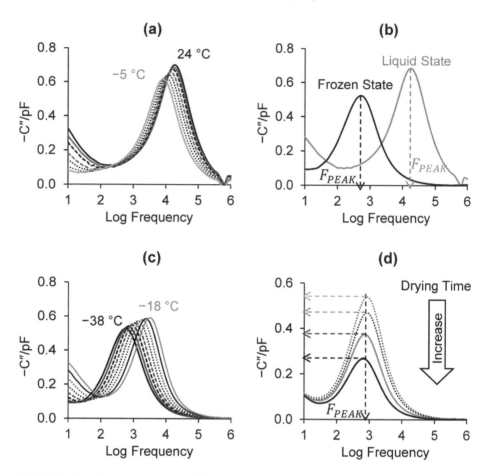

FIGURE 5.3 Basic responses of TVIS dielectric loss spectrum of double distillation water, 3.5 g, contained in a type I tubular glass vial with a pair of 10 × 19 mm electrodes attached to the external surface of vial during freeze-drying process: (a) cooling the liquid phase, (b) phase transition (liquid water to ice), (c) heating of the solid phase, and (d) primary drying.

dependence of the electrical resistance of the water. However, when frozen, the shift in the characteristic frequency towards lower frequencies (Figure 5.3b) is due to the formation of ice (which itself has a dielectric relaxation) rather than the increase in resistance associated with the formation of the solid phase. The electrical resistance of the sample is now much higher than in the liquid state, so we expect that the Maxwell-Wagner polarization of the interface between the glass wall and the ice will be shifted to frequencies below the lower limit of the experimental frequency window.

The dielectric relaxation time of ice increases as the temperature is reduced and so the peak shifts to lower frequencies. The peak frequency (F_{PEAK}) therefore provides a surrogate measurement for the ice temperature (Figure 5.3c). The calibration of this TVIS parameter for temperature may be achieved by introducing a temperature cycling stage (of reheating and recooling) once the ice has been frozen from the solution.

The height of the relaxation peak (C''_{PEAK}) is proportional to the height of the ice cylinder in direct contact with the inside of the glass wall. For a cylindrical ice mass (where the volume and therefore mass is proportional to the ice cylinder height), it follows that the TVIS parameter C''_{PEAK} may be calibrated for the ice mass to provide an opportunity to measure the drying rate in primary drying (Figure 5.3d). The caveat, however, is that in both cases of temperature and ice mass predictions during primary drying (using electrodes attached to the wall of the vial) the ice cylinder (which is defined in part by the shape of the sublimation interface) must not change shape as the ice cylinder reduces during the sublimation process. This condition is met for the freeze-drying of ice formed from solutions, and especially for those core vials at the center of the shelf, but only holds true for a limited period of time for the freeze-drying of ice that is formed from pure water. This requirement is likely to be less important in future applications of TVIS where the electrodes are removed from the vial and replaced by a single electrode placed above the vial(s), which then measures through the vial base to the freeze-dryer shelf (the ground plane).

In this chapter we devote our attention to the application of TVIS to

1. the calibration of the liquid state and solid-state temperatures
2. the identification of the two critical events in the freezing stage of the lyophilization cycle, that is, the determination of the ice nucleation temperature and the end of solidification phase

It should be recognized that the ice nucleation temperature and the solidification end point may in fact be determined in one relatively straightforward manner, by using a thermocouple placed within the vial. And so that begs the question as to whether there is a need for an alternative technology for this purpose. The reason for an alternative technology is that the very presence of the thermocouple alters the way in which the ice freezes and therefore changes the ice crystal structures (Roy and Pikal 1989). It follows that the thermocouple containing vial is not representative of those vials in its locality during the freezing stage. This is equally true for the primary drying stage whereby the drying rate is affected by the altered ice structures (which impact the vapour flow resistance of the dry layer) and the additional heat source provided by the thermocouple.

5.3   TEMPERATURE CALIBRATION OF THE TVIS SYSTEM

First, we shall concentrate on methods for calibrating the TVIS response for the temperature of the liquid state (prior to the onset of ice formation, i.e., ice nucleation).

An obvious way to approach the calibration of the TVIS response for temperature would be to place a temperature sensor such as a thermocouple inside the vial (notwithstanding what has just been said about the impact of the sensor on ice crystal structures). Unfortunately, because the metal of a hardwired thermocouple (as opposed to those temperature sensors that are operated wirelessly) is electrically grounded, it introduces a significant distortion in the impedance spectrum if one attempts to measure the TVIS response while the thermocouple remains inside the vial and in contact with the liquid (Figure 5.4).

We note here that this perturbation is much less for ice or a frozen solution.

There are a number of approaches that might address this issue:

1. the use of temperature sensors in nearest-neighbour vials to infer the product temperature in the test vial
2. the use of nonmetallic sensors (based on fiber optic technology) or wireless sensors (such as Tempris®)
3. the use of planar thermocouples attached to the outside of the vial (Parvis et al. 2012, 2014; Grassini et al. 2013)

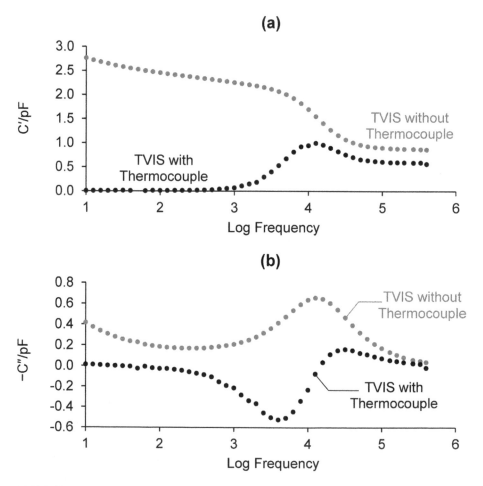

FIGURE 5.4 (a) Real-part spectra and (b) imaginary-part spectra showing the impact on the TVIS spectrum of a thermocouple placed within a TVIS vial containing ultrapure water relative to a vial without a thermocouple.

Here we describe the first two of these approaches:

The first approach is to use temperature measurements in the nearest-neighbour vials (i.e., those vial in close proximity of the TVIS vial and at a height that corresponds to the center point of the ice layer in the region bounded by the electrode) in order to predict the product temperature in the TVIS vial (Figure 5.5).

The thermocouples used should comply with best practices for wired sensors as defined by Nail et al. (2017). The disadvantage of the approach is that there is invariaby a distribution in temperatures between vials, and therefore this approach will inevitably introduce some uncertainty in the calibration of F_{PEAK}. This uncertainty can be determined by taking half of the range of temperatures recorded in a minumum of three of the nearest-neighbour vials. Typically this range is in the region of 0.4°C, so an uncertainty limit might be in the region of ±0.2°C. The advantage, however, is that it avoids having an invasive temperature sensor in the vial that provides another heat source and additional nucleation sites that would otherwise impact the way the ice freezes and dries.

The second approach is to place a nonmetallic element, such as a fibre optic probe with a Bragg grating or a wireless temperature sensor based on a quartz crystal resonator (Tempris®), directly in the TVIS test vial. Fibre optic sensors have been used previously in freeze-drying applications (Friess et al. 2009; Kasper et al. 2013; Horn and Friess 2018) and are shown in our studies to be nonperturbing to the TVIS spectrum. Wireless sensors (Schneid and Gieseler 2008) are becoming commonplace in freeze-drying production applications and in our studies are also shown to be nonperturbing to the TVIS spectrum (Figure 5.6).

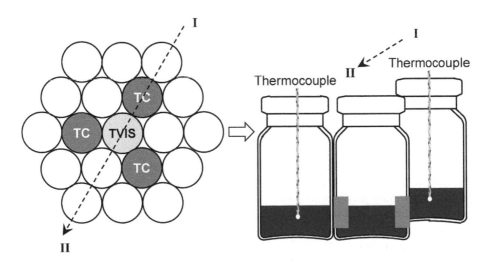

FIGURE 5.5 Triangulation method for temperature calibration of the TVIS vial showing a schematic of the position of three thermocouple-containing vials (marked TC) as nearest neighbours for the TVIS vial (marked TVIS). The triangulation of temperature measurements may be used to infer the temperature of the material inside the TVIS vial to an accuracy that is approximately equal to the deviation from the mean thermocouple temperature at any point in time during the reheating of the population of vials.

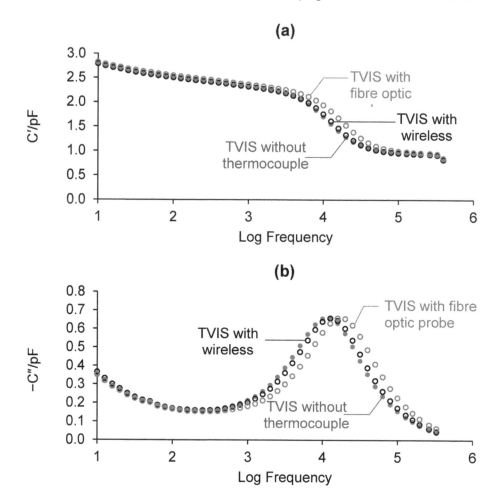

FIGURE 5.6 The similarity in shape of real-part spectra (a) and imaginary-part spectra (b) of ultrapure water between a TVIS vial with a fibre optic sensor, a wireless sensor (Tempris®), and the same vial without any temperature sensors.

By modeling (Smith et al. 2017) it has been demonstrated that log F_{PEAK} is dependent on the average temperature in the region bounded by the electrodes (assuming a linear temperature gradient across this region). Therefore in both cases of the triangulation method using nearest-neighbour vials containing thermocouples (to interpolate the temperature in the TVIS vial) or the direct method with the fiber optic sensor or wireless sensor inside the TVIS vial, the tips of the sensors should be positioned at a height that is defined by the midpoint of the volume of liquid or frozen solid in the region bounded by the electrodes (Figure 5.7). This height is given by $d_m + d_b$.

5.3.1 Liquid Temperature Calibration

The first application for this temperature calibration would be to exploit the temperature dependence of the Maxwell-Wagner process that is observed for the liquid-filled vial. This works well for water and for solutions of neutral solutes such as the sugars, as the

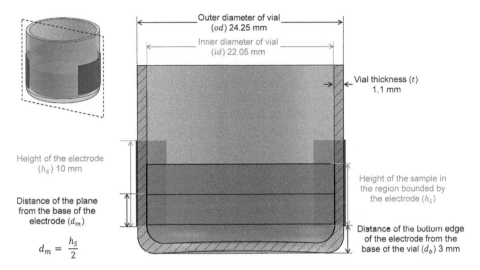

FIGURE 5.7 Illustration of a TVIS vial showing the dimensions. d_b, the distance of the bottom edge of the electrode from the base of the vial; h_s, height of the sample in the region bounded by the electrode; h_E, height of the electrode; d_m, distance of the midpoint of the sample contained within the region bounded by the electrodes from the base of the vial.

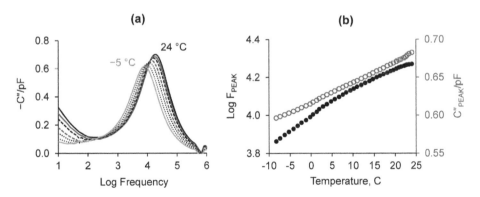

FIGURE 5.8 (a) Imaginary capacitance spectra of double-distilled water inside a TVIS measurement vial (10 ml nominal capacity) with a pair of 10 × 19 mm electrodes (positioned with the lower edge of the electrode at a height of 3 mm from the base of the vial). (a), The temperatures for the selected spectra are inferred from thermocouple measurements in nearest-neighbour vials (according to the triangulation method described in the text). (b), Temperature dependencies of the TVIS parameters log F_{PEAK} (filled circle) and C''_{PEAK} (open circle) associated with the supercooling of water in the TVIS vial.

relaxation frequency of the MW process is well within the experimental frequency range of the current TVIS technology. Both F_{PEAK} and C''_{PEAK} have temperature dependencies of comparable magnitude, and so in theory either parameter could be used. However, given that C''_{PEAK} also depends on the height of the liquid in the vial then it is preferable to use F_{PEAK} for temperature calibration for water in the liquid state (Figure 5.8).

Temperature calibration for the liquid state can be incorporated within a conventional cycle, which inevitably starts with the cooling phase in order to reduce the liquid temperature to some sub-zero temperature (super-cooling) and initiate ice nucleation. For the purpose of the calibration, TVIS spectra are recorded (e.g., every 2 min) simultaneously with the recording of the temperatures from the sensors in the nearest-neighbour vials as the liquid cools. It is convenient that the cooling rates employed in a conventional cycle (0.5 to 1°C min⁻¹) are sufficiently slow that some degree of thermal equivalence in maintained between the TVIS vial and the nearest-neighbour vials. Nevertheless, there will be a degree of uncertainty in the temperature predicted from the nearest-neighbour vials which is comparable with the temperature variation between those vials being used for calibration purposes. As stated earlier, this variation is typically in the region of ±0.2°C and is then taken as our estimate for the uncertainty in any temperature inferred from this calibration approach, that is, in the determination of the nucleation temperature in the freezing stage.

5.3.2 ICE TEMPERATURE CALIBRATION

The sensitivity of the peak frequency (F_{PEAK}) to the temperature of the ice in the volume space bounded by the external electrodes is largely due to the temperature dependency of the dielectric relaxation time of ice. The possibility exists to calibrate this response against known product temperatures (as determined by thermocouple or RTD sensors) by the inclusion of a thermal cycling stage after the freezing stage. Here it is the calibration coefficients for the reheating phase of the thermal cycle (rather than the recooling phase) that are used to predict the temperature in primary drying (Figure 5.9).

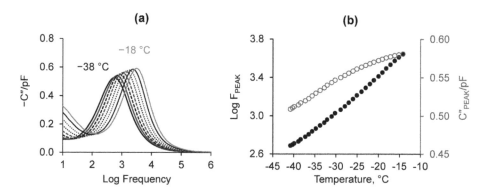

FIGURE 5.9   (a), Imaginary capacitance spectra of frozen water contained in a 10 mL type I tubular glass vial with a pair of 10 × 19 mm electrodes (positioned with the lower edge of the electrode at a height of 3 mm from the base of the vial), demonstrating the temperature dependency of the dielectric loss peak of ice during the reheating phase of a thermal cycle. The temperatures for example spectrum are interpolated from thermocouple measurements in nearest-neighbour vials. (b), Temperature dependencies of the TVIS parameters log F_{PEAK} (filled circle) and $C_{PEAK}^{''}$ (open circle) of ice during reheating from −40 to −10°C with 0.5°C min⁻¹ rate.

Temperature calibration is achieved by having frozen the product (or water) to between −50 and −45°C, and then holding for 1–2 h before reheating at a slow rate (e.g., 0.5°C min⁻¹) whilst recording both the TVIS spectra and the temperature from the sensing probes. Unlike for the liquid state, for which the calibration can be undertaken as an inherent part of the cycle, the solid-state calibration requires the incorporation of an additional thermal cycling stage.

In one case it might be that the cycle demands an annealing stage to 'ripen' the ice crystal structures and so the product temperature is taken from the solidification temperature to a temperature above T_g' but below the melting temperature before returning product temperature to a value below T_g', prior to application of the vacuum (Figure 5.10a).

Given that the impedance of the product is likely to change as a result of annealing, it will be necessary to include an additional reheating stage that takes the product temperature back to the glass transition in order to capture the calibration

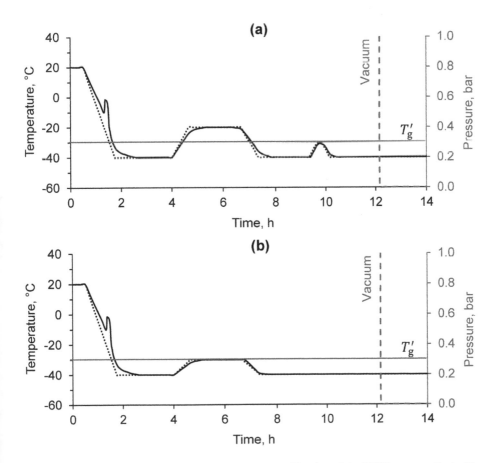

FIGURE 5.10 Examples of thermal cycles for the calibration of the TVIS system. Cycle (a) applies to the condition when the product is expected to be annealed and cycle (b) applies to the condition when the product is not supposed to be annealed. In both cases the solid black line represents the product temperature and the dotted line represents the shelf temperature.

coefficients for the TVIS parameter, F_{PEAK}, before almost immediately returning it to the target initial temperature prior to the application of the vacuum (Figure 5.10a). It should be noted that, before application of the vacuum, the product temperature will be higher than the shelf temperature (by a few degrees centigrade) and therefore this should be taken into account when setting the shelf temperature. In other words, the shelf temperature should be a few degrees below the target product temperature (i.e., T_g') in order to ensure that the product doesn't anneal.

In another case it might be that the product should not be annealed in order to protect the stability of the active pharmaceutical ingredient (API), and so the product temperature is taken from the solidification temperature to a temperature below T_g' and then back to the target temperature prior to the application of the vacuum (Figure 5.10b). Given that a temperature cycle which maintains the product below T_g' is unlikely to have annealed the product (i.e., change its ice structure and hence its impedance), then it is expected that the calibration coefficients may be captured during the first reheating stage without any further recycling of the temperature.

In both cases, the initial equilibrium product temperature in the initial stages of primary drying (prior to the development of the dry layer resistance) will be defined by the chamber pressure, and hence the lower limit of product temperature must at least be captured in the range of temperatures programmed for the thermal cycle/calibration.

Here we use a simple polynomial function to model the temperature dependency of the TVIS peak frequency associated with the dielectric relaxation of the ice. However, a more rigorous approach might be to fit the data with a Cole-Cole relaxation function in order to determine the relaxation time and then to use the equations (15 and 16) described by Popov et al. (2015, 2017) to model the characteristics (shape) of the calibration curve; a model which takes into account the cross-over behaviour in the polarization mechanism of ice from orientation defect propagation at higher temperature (>235 K) to ionic defect propagation at lower temperature (<235 K).

A typical relationship between the dielectric relaxation peak frequency and the temperature of ice is shown in Figure 5.11.

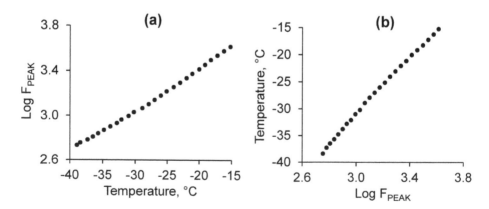

FIGURE 5.11   (a), Temperature coefficient and (b) temperature calibration plots of log F_{PEAK} for the ice relaxation peak vs the temperature in the nearest-neighbour vial. The object under test is a 10 mL glass tubing vial with a pair of 10×19 mm electrodes at a distance of 3 mm from the base of the vial, containing 3.5 g double-distilled water.

TABLE 5.1
Temperature Calibration Parameters and Temperature Coefficients of the Dielectric Relaxation of Ice within the TVIS Vial

Order of Fitting Coefficient	2	1	0
Temperature Calibration	−4.73E+00	−5.68E+01	−1.59E+02
Temperature Coefficient	2.50E−04	5.07E−02	4.33E+00

(Note: The temperature calibration parameters provide the means to predict temperature from log F_{PEAK} in the subsequent drying step, whereas the temperature coefficient defines the dependency of log F_{PEAK} on temperature.)

The fitting of polynomial functions to (1) the plot of log F_{PEAK} vs temperature and (2) the plot of temperature vs log F_{PEAK} (Table 5.1) is used to generate a set of temperature coefficients that demonstrate the temperature dependence of F_{PEAK} parameter and a set of calibration coefficients that are expected to provide the basis for the prediction of the product temperature during primary drying.

Here, we introduce the parameter $T_{(F_{PEAK})}$ to define those temperatures that predicted during the subsequent primary drying stage from the F_{PEAK} calibration that was recorded during the annealing stage.

5.4 ICE FORMATION (NUCLEATION AND GROWTH)

Freeze-drying is a multistage process that starts with the conversion of the majority of the liquid water to ice by supercooling the solution below the solution melting point. The onset of ice nucleation, or rather the temperature at which the ice nucleates (T_n), is important to understand how the ice structure develops, as it defines the size distribution of the interconnected pores and hence the porosity of the dry layer that is left behind once the ice has sublimed:

1. Slow freezing gives time for nucleation to occur at higher temperatures, by providing an extended period for ice nuclei to form around surfaces and foreign objects (particles in the solution). Nucleation at higher temperatures then facilitates growth by maintaining water molecule mobility and hence diffusion to the ice front, and therefore the ice crystals that form are generally of a larger size, which provides a more interconnected network of larger channels and therefore a lower dry layer resistance.
2. Fast freezing pushes the product to low temperature more rapidly, and hence nucleation and then ice crystal growth tends to happen at lower temperatures, which then reduces growth rates through a reduction in the rate of water molecule diffusion to the ice front, resulting in smaller ice crystals and a higher dry layer resistance.

It follows that the nucleation temperature is one of the critical process parameters in the design space (Arsiccio and Pisano 2018). However, the stochastic nature of the process means that the nucleation onset temperature is difficult to control, unless additional technologies, such as the controlled nucleation technologies of pressure shock (Konstantinidis et al. 2011) and ice fog (Rambhatla et al. 2004), are deployed within the dryer. This will be discussed in more detail in Chapter 6. It is also important to note that in the sterile, particle-free environment of a commercial scale freeze-dryer and with an aseptically filtered product one expects different nucleation behaviour from the typical laboratory research tests within a less controlled environment.

It is a standard practice to measure the ice nucleation temperature with a product probe (i.e., an RTD or thermocouple inserted in a test vial containing the product). However, the probe itself alters the nucleation event by providing additional surfaces around which the nucleation event can occur while impacting the temperature in the vial, as it provides an additional heat source. In effect the probe alters not only the process one is trying to measure (i.e., the nucleation onset temperature) but it will also impact the drying rate from that particular vial.

Photographic images of the nucleation event within a thermocouple vial and a TVIS vial are show in Figure 5.12.

In this particular example, it is clear that the ice is forming around the thermocouple, whereas in the TVIS vial the ice forms in the traditional manner of ice forming across the internal surface area of the vial.

An example TVIS study of the ice nucleation and growth phase is shown in Figure 5.13.

Each graph focusses on a 2 hour time period during which the product temperature is reduced from ~0°C to −40°C via a linear ramp of 0.5°C min⁻¹. The three-dimensional plots of real and imaginary capacitance vs frequency and time (Figure 5.13a and 5.13e) capture the entire response surface and provide a first indication of what information might be extracted in terms of TVIS process parameters. These are the peak frequency (F_{PEAK}) and peak amplitude (C''_{PEAK}) from the imaginary capacitance spectra (Figure 5.13a) and the values for the real-part capacitance at the limits of low and either side of the relaxation process. Here, we illustrate only the high frequency limit of the real-part capacitance by selecting the time line for the 0.2 MHz frequency point (Figure 5.13e). We refer to this parameter as C'(0.2 MHz), where the number in brackets signifies the experimental frequency chosen for analysis.

FIGURE 5.12 Ice forming occurring in a 10 mL type 1 glass tubing vial with thermocouple inside (left) and the impedance measuring vial (right) with an electrode pair (each 10 mm high and 19 mm wide, separated from the base by a 3 mm gap).

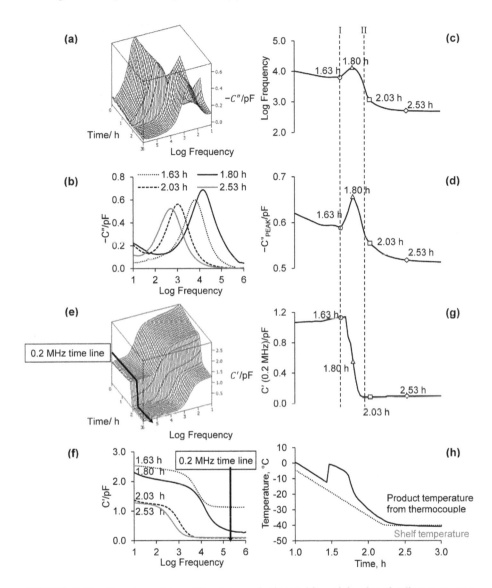

FIGURE 5.13 Freezing of two 10 mL type 1 glass-tubing vials placed adjacent to one another with each containing 3.5 g of double-distilled water. One has a pair of 10 × 19 mm electrodes attached to the outside wall of the vial at a height of 3 mm from the base. The other has a thermocouple immersed in the water. (a) and (e) are the response surfaces of the imaginary capacitance (dielectric loss) and real-part capacitance (dielectric storage) as a function of frequency and time; (b) and (f) are the real and imaginary capacitance spectra at selected time points; (c) and (d) are the time profiles of the TVIS parameters F_{PEAK} and C''_{PEAK} ; (g) is the time profile of the TVIS parameter C' (0.2 MHz); (h) is the time profile of the thermocouple temperature. The ice nucleation event (at time point I) is reflected in the time profiles of both F_{PEAK} and C''_{PEAK}, whereas the solidification end point (II) is reflected in the inflection in the real-part capacitance measured at high frequency, C' (0.2 MHz).

These parameters are then plotted as a function of time during the freezing stage, in order to establish which are sensitive to the onset of ice formation (nucleation) and which might be sensitive to the end of the ice solidification phase.

From a close examination of Figure 5.13 it appears that the parameters of the TVIS response for water/ice that are most sensitive to the nucleation onset point are F_{PEAK} and C''_{PEAK}. At first glance, both appear to reflect the spike in temperature that is associated with the exothermic process of ice formation. However, the dielectric mechanisms responsible for these spikes are likely to be a complex function of the heterogeneous structures that coexist during this phase transition as well as the temperature dependencies of each structural domain. Some thoughts on these dielectric mechanisms are suggested here.

Given that ice forms from the bottom of the vial to the top of the vial, then for at least part of the time period during which ice forms, there will be a residual liquid volume (albeit shrinking) on top of the ever-increasing volume of the ice phase. The liquid that remains at any particular time can then provide in-part one of the mechanisms underpinning the spike in each parameter, through the positive temperature dependency of the relaxation frequency and capacitance increment of the Maxwell-Wagner process associated with the liquid phase. However, that alone cannot be wholly responsible for the spikes in F_{PEAK} and C''_{PEAK} because the maximal increase in both F_{PEAK} and C''_{PEAK} exceeds the initial value at 0°C and so there must be other more dominant mechanisms at play. One tentative suggestion is that the creation of intermediate ice structures prior to the completion of solidification across the liquid creates opportunities for further interfacial polarization effects within the bulk ice mass that collectively have higher dipole moments than the completely frozen ice mass.

For the case in which the TVIS vial nucleates before the thermocouple-containing vial then the nucleation temperature in the TVIS vial can be 'read' simply from the average temperatures in the nearest-neighbour vials. However, if the TVIS vial doesn't nucleate first, then as soon as ice begins to nucleate in the nearest-neighbour thermocouple vials, the resultant spike in temperature will inevitably preclude any further direct prediction of the temperature in the TVIS vial. In this case, the nucleation temperature in the TVIS vial may not be inferred directly from the thermocouple-containing vials. However, by fitting a suitable curve to the plot of log F_{PEAK} against the average temperature from thermocouple vials (Figure 5.14) then it is possible to extrapolate the curve to the nucleation point (i.e., when log F_{PEAK} starts to increase dramatically) in order to predict the nucleation temperature, T_n.

Whether the TVIS vial nucleates first or second, the fact that the TVIS vial doesn't contain an invasive thermocouple means that one can then consider the TVIS vial to be representative of vials placed at that location in the dryer in terms of the ice crystal structure that forms on freezing and the subsequent sublimation characteristics in the primary drying stage.

Having described a noninvasive approach for the determination of the nucleation temperature we now describe an approach for the determination of the solidification end point, one that does not rely on any changes in the temperature of the contents of the vial. The reason why temperature is not ideal for the assessment of the solidification end point can be explained as follows. The freezing process, in many conventional cycles, comprises a slow ramp from a temperature above the freezing

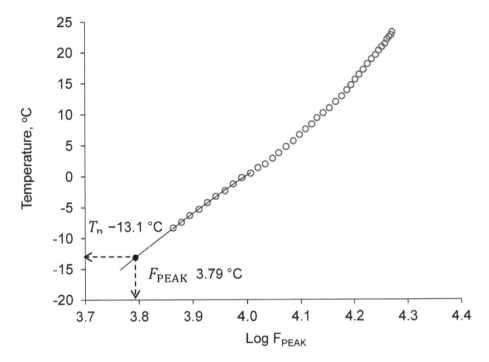

FIGURE 5.14 Prediction of ice nucleation temperature (T_n) by using a calibration plot between log F_{PEAK} and thermocouple temperature of ice within an adjacent vial.

point to the solidification temperature. It is therefore likely that the solidification of ice is completed, in advance of the shelf temperature reaching its set value. Using a temperature sensor, such as a thermocouple, to predict the solidification end point is complicated by the fact that the temperature in the vial continues to change beyond the solidification end point, and so the end point is not so clear.

Here we exploit the fact that the high-frequency real-part capacitance of ice has a very low temperature coefficient (i.e., it does not change greatly with fluctuations in temperature). It follows that once all the ice has formed then the high frequency capacitance of the TVIS vial settles quickly to a value that remains almost constant, despite the fact that the temperature of the vial may continue to change as the system attempts to lose the excess latent heat that was generated on formation of the ice. This provides a more distinct end point for the completion of the solidification phase.

With the current instrument that was developed for TVIS measurements, and which has provided all the data for this chapter, there is some distortion to the high frequency end of the spectrum, and so we have used the 200 kHz frequency point to determine the solidification end point. Figure 5.13e shows the response surface of the real-part capacitance as a function of frequency and time. The solid black line highlights the time line for the 200 kHz frequency, which is then shown in Figure 5.13g. From Figure 5.13g one can easily recognize the solidification end point from the inflection where the steep descent in the real-part capacitance transitions to an almost constant value (see vertical line II).

5.5 CONCLUSIONS

This chapter describes opportunities for using TVIS in the noninvasive measurement of the ice nucleation temperature and the end point of ice formation (complete solidification). The sensitivity of the TVIS dielectric loss peak to the formation of ice is due to the transition in the mechanism of relaxation from the Maxwell-Wagner polarization of the glass wall (through the electrical resistance of the liquid water phase) to the dielectric relaxation of ice. During the crystallization of ice, it is inevitable that both the liquid and solid state will coexist during the period of time from the onset of ice formulation (nucleation) to the complete solidification of ice. Fortunately, the relaxation frequencies of both processes have positive temperature dependencies, and so, regardless of how much of the water is in either of the liquid and solid states, one can expect the peak frequency to be sensitive to the spike in temperature that originates from the release of latent heat, as the liquid transitions to the solid state (ice). This in effect underpins the use of the TVIS parameters in the determination of the nucleation onset time.

For the end point of ice formation, we have proposed an alternative approach to that based on the observation of a temperature spike, which instead uses the temperature-independent high-frequency capacitance of ice.

In future publications we will extend this work to include the ice nucleation temperature and end points of the solidification of ice that forms from aqueous solutions. The issue to be addressed then, with using the dielectric loss peak for monitoring the ice formation process in conductive samples (such as a protein solution, a drug solution, or even a sugar solution with added salt) is that the peak frequency will be outside the frequency range of the instrument when the sample is in the initial liquid state. This is because the dielectric loss peak for the liquid state material is due to a Maxwell-Wagner (interfacial) polarization of the glass wall through the solution resistance, and if the resistance of the solution is low then the loss peak frequency shifts to high frequency and the dielectric loss parameter then becomes less sensitive to any changes in the properties of the solution inside the vial. An alternative approach is therefore required.

ACKNOWLEDGMENTS

The original TVIS system used to generate the spectra within this book chapter was developed by Evgeny Polygalov and Geoff Smith (from De Montfort University) in a collaboration with GEA Pharma Systems (Eastleigh, UK) and was part-funded by a UK government, Innovate UK Collaborative R&D project called LyoDEA (project reference: 10052).

REFERENCES

Arshad, M. S., Smith, G., Polygalov, E., and I. Ermolina. 2014. Through-vial impedance spectroscopy of critical events during the freezing stage of the lyophilization cycle: The example of the impact of sucrose on the crystallization of mannitol. *Eur. J. Pharm. Biopharm.* 87:598–605.

Arsiccio, A., and R. Pisano. 2018. Application of the quality by design approach to the freezing step of freeze-drying: Building the design space. *J. Pharm. Sci.* 107:1586–1596.

Careri, G., Geraci, M., Giansanti, A., and J. A. Rupley. 1985. Protonic conductivity of hydrated lysozyme powders at megahertz frequencies. *Proc. Nat. Acad. Sci.* 82:5342–5346.

Friess, W., Resch, M., and M. Wiggenhorn. 2009. Monitoring device for a dryer. U. S. Patent Application US20110247234A1, filed December 29, 2009.

Grassini, S., Parvis, M., and A. A. Barresi. 2013. Inert thermocouple with nanometric thickness for lyophilization monitoring. *IEEE Trans. Instrum. Meas.* 62:1276–1283.

Horn, J., and W. Friess. 2018. Detection of collapse and crystallization of saccharide, protein and mannitol formulations by optical fibers in lyophilization. *Front. Chem.* 6.

Kasper, J. C., Wiggenhorn, M., Resch, M., and W. Friess. 2013. Implementation and evaluation of an optical fiber system as novel process monitoring tool during lyophilization. *Eur. J. Pharm. Biopharm.* 83:449–459.

Konstantinidis, A. K., Kuu, W., Otten, L., Nail, S. L., and R. R. Sever. 2011. Controlled nucleation in freeze-drying: Effects on pore size in the dried product layer, mass transfer resistance, and primary drying rate. *J. Pharm. Sci.* 100:3453–3470.

Nail, S. L., Tchessalov, S., Shalaev, E., et al. 2017. Recommended best practices for process monitoring instrumentation in pharmaceutical freeze drying – 2017. *AAPS PharmSciTech* 18:2379–2393.

Parvis, M., Grassini, S., and A. A. Barresi. 2012. Sputtered thermocouple for lyophilization monitoring. Proceedings of Instrumentation and Measurement Technology Conference (I2MTC-2012), May 13–16, 2012, pp. 1994–1998.

Parvis, M., Grassini, S., Fulginiti, D., Pisano, R., and A. A. Barresi. 2014. Sputtered thermocouple array for vial temperature mapping. Proceedings of Instrumentation and Measurement Technology Conference (I2MTC-2014), May 12–14, 2014, pp. 1465–1470.

Popov, I., Lunev, I., Khamzin, A., Greenbaum, A., Gusev, Y., and Y. Feldman. 2017. The low-temperature dynamic crossover in the dielectric relaxation of ice I H. *Phys. Chem. Chem. Phys* 19:28610–28620.

Popov, I., Puzenko, A., Khamzin, A., and Y. Feldman. 2015. The dynamic crossover in dielectric relaxation behavior of ice I H. *Phys. Chem. Chem. Phys.* 17:1489–1497.

Rambhatla, S., Ramot, R., Bhugra, C., and M. J. Pikal. 2004. Heat and mass transfer scale-up issues during freeze drying: II. Control and characterization of the degree of supercooling. *AAPS PharmSciTech* 5:54–62.

Roy, M. L., and M. J. Pikal. 1989. Process control in freeze drying: Determination of the end point of sublimation drying by an electronic moisture sensor. *J. Parent. Sci. Tech.* 43:60–66.

Schneid, S., and H. Gieseler. 2008. Evaluation of a new wireless temperature remote interrogation system (TEMPRIS) to measure product temperature during freeze drying. *AAPS PharmSciTech* 9:729–739.

Smith, G., and E. Polygalov. 2019. Through-vial impedance spectroscopy (TVIS): A novel approach to process understanding for freeze-drying cycle development. In *Lyophilization of pharmaceuticals, vaccines and medical diagnostics*, ed. K. Ward, and P. Matejtschuk. New York: Springer.

Smith, G., Arshad, M. S., Polygalov, E., and I. Ermolina. 2013. An application for impedance spectroscopy in the characterisation of the glass transition during the lyophilization cycle: The example of a 10% w/v maltodextrin solution. *Eur. J. Pharm. Biopharm.* 85:1130–1140.

Smith, G., Arshad, M. S., Polygalov, E., and I. Ermolina. 2014. Through-vial impedance spectroscopy of the mechanisms of annealing in the freeze-drying of maltodextrin: The impact of annealing hold time and temperature on the primary drying rate. *J. Pharm. Sci.* 103:1799–1810.

Smith, G., Arshad, M. S., Polygalov, E., Ermolina, I., McCoy, T. R., and P. Matejtschuk. 2017. Process understanding in freeze-drying cycle development: Applications for through-vial impedance spectroscopy (TVIS) in mini-pilot studies. *J. Pharm. Innov.* 12:26–40.

Smith, G., Jeeraruangrattana, Y., and I. Ermolina. 2018. The application of dual-electrode through vial impedance spectroscopy for the determination of ice interface temperatures, primary drying rate and vial heat transfer coefficient in lyophilization process development. *Eur. J. Pharm. Biopharm.* 130:224–235.

Suherman, P. M. 2001. A novel dielectric technique for monitoring the lyophilisation of globular proteins. PhD diss., De Montfort University.

Suherman, P. M., Taylor, P. M., and G. Smith. 2002. Development of a remote electrode system for monitoring the water content of materials inside a glass vial. *Pharm. Res.* 19:337–344.

6 Innovations in Freeze-Drying Control and In-Line Optimization

Antonello Barresi, Roberto Pisano, and Davide Fissore

CONTENTS

6.1 INTRODUCTION

As discussed in the book introduction, freeze-drying is considered a "gentle process", but the product temperature must be maintained below a threshold value (denaturation or scorch temperature), that is characteristic of the product being processed. In the case of freeze-drying of liquid solutions containing a drug (and one or more excipients), for a crystallizing product a further constraint is represented by the eutectic temperature of the system, to avoid product melting. In the case of amorphous products, the collapse temperature must not be trespassed, to avoid the collapse of the dried cake which also causes higher reconstitution times, higher amount of water in the final product, etc. It must be considered, anyway, that if the product is partially crystalline, temperatures higher than the collapse temperature can be adopted, and the robustness of the cycle must be evaluated (Tchessalov and Warne 2008).

The operating conditions of the freeze-drying process do not influence only product temperature but also the sublimation flow rate and, finally, the drying time. The requested vapor flow rate must be compatible with the capacity of the condenser, and with the conductance of the duct and isolating valve, to avoid the occurrence of choking flow and to allow maintenance of the desired pressure in the chamber. If the flow becomes sonic in the duct, pressure control is lost, and the undesired pressure increase in the chamber causes product overheating. A proper design of the equipment is required, as will be discussed in Chapter 9.

The modern concept of closed-loop control of the critical quality attributes of the product is rarely adopted in the pharmaceutical field, especially in industrial units. It was introduced for freeze-dryers in the early 1960s, but until now has been generally only an automatic sequence of operations, with predefined set point values, regulating shelf temperature and chamber pressure to maintain them within the required tolerance. Thus, many of the control strategies proposed in the patent and scientific literature are still valid today, and their implementation in pilot and production units would represent a strong advancement in current technology (Nail and Gatlin 1985; Jennings 1999)

Things have been slightly changing in the last years, moving toward a product quality "built in or by design" instead of limiting the test to the final product quality, following the "Guidance for Industry PAT—A Framework for Innovative Pharmaceutical Manufacturing and Quality Assurance" first issued in 2004 (Food and Drug Administration 2004).

The use of an active control system allows for identification of potential disturbances or the approach of critical conditions (mass transfer limitation, choked flow, etc.) and modifies the operating parameters to react to them, adapting also to different input conditions. This would not only help to avoid failures but also to find in-line optimal conditions, minimizing the drying time without impairing the product quality. It must be stressed that if the product is thermally sensitive, for example, the final quality is not guaranteed by ensuring that the shelf temperature is that fixed in the developing stage (in a different equipment, with different sensors, different loading, and possibly different container and cake structure, caused by different freezing conditions), but that the product temperature remains in the desired interval. Such an automatic tool might be very useful also for cycle development, because it would be enough to set the limit or target temperature of the product, to obtain automatically the optimized cycle, according to the constraints inserted.

It must be said that batch manufacturing pharmaceutical products poses a lot of constraints (e.g., sterility conditions, compatibility with automatic loading/unloading, the necessity to stopper vials at the end of the process). One of the main technical limits of the application of closed-loop control systems in freeze-drying is the difficulty in measuring the temperature of the product, as inserting a probe is often not possible in production and gives an unreliable or un representative measure (as discussed in Chapter 4). In addition, there is no way with current technology to measure directly in production the advancement of drying (related to the position of the ice–dry product interface), to know when to stop primary drying. Thus, on the one hand, in-line control and optimization can be improved only by improving monitoring of the process, with new PAT tools that allow measurement or inference of the product temperature and possibly also its variance in the batch. On the other hand, it is fundamental to evaluate in-line the most important parameters of the process, with process identification instruments, and to adopt a model-based approach to predict product evolution and eventually to estimate the final product quality (Fissore et al. 2009b, 2012a, 2012b).

The PAT tools currently available for monitoring the freeze-drying process have been described in detail elsewhere (Barresi et al. 2009a; Johnson et al. 2009; Patel and Pikal 2009; De Beer et al. 2009, 2011; Patel et al. 2010; Barresi and Fissore 2011; Rosas et al. 2014; Nail et al. 2017; Fissore et al. 2018), while advanced control was discussed in a previous book (Barresi et al. 2018). Infrared imaging and its application to control have been discussed in Chapter 4.

The aim of this chapter is to discuss open- and closed-loop control systems applied to a freeze-drying process, evidencing the possibility of model-based approaches for rejection of disturbances, failure prevention, and process optimization. At the end, it will be shown how some of the recently developed tools can be useful for automatic cycle development. All three steps of the process, namely freezing, primary drying, and secondary drying, will be analyzed, with emphasis on how control of the freezing step can be beneficial and interact with the optimization of the whole process. Extension of monitoring and control to continuous processes will be discussed in Chapter 8.

6.2   OPEN- AND CLOSED-LOOP CONTROL IN FREEZE-DRYING

6.2.1   CONTROL STRATEGIES

The most commonly measured variable in the freeze-drying process control is the product temperature. The insertion of probes is not generally allowed in production apparatus, and thus, their use is generally limited to pilot and lab scale in the development stage, even if wireless probes and smart-vials with sputtered thermocouples and thermocouple arrays have been recently proposed (Schneid and Gieseler 2008; Bosca et al. 2013a; Corbellini et al. 2010; Grassini et al. 2013; Oddone et al. 2015). Methods based on process response curves, like the barometric temperature measurement (BTM) or the pressure rise test (PRT), are more suitable for industrial equipment, as will be discussed in the next section. As an alternative, it is possible to monitor the sublimation flux (which can also be related to the product temperature with some assumptions and simplifications), and some methods have been recently proposed and validated also for control applications (Gieseler et al. 2007a; Chen et al. 2008; Vollrath et al. 2017).

As concerns the manipulated variables, they can be only the shelf temperature and/or the chamber pressure, from which the heat flux from the shelf to the product depends.

The common practice is to manipulate the shelf temperature to control the product temperature, even if the performance of the method is not very good, because heat transfer control by this way is slow, owing to the thermal inertia of the system. To take into account the nonhomogeneity of the batch, it was proposed that the temperature in several vials be measured. Rey (1976) proposed to manipulate the shelf temperature to maintain as constant the resistivity of the product; the control loop proposed may be unstable because of the large lag of shelf heating and cooling, compared to the rapid variation of resistivity with temperature. Additionally, the measure of the product's conductivity is very invasive and used principally at lab scale in the process development stage, when there is a risk of eutectic melting.

A simple control system that uses the ice temperature measured with the BTM method to manipulate the shelf temperature was also proposed a long time ago, showing that the temperature of ice can be considered constant within ±1°C by using a quickly reacting heating system and that the automatic control system could be built in an industrial unit (Oetjen et al. 1962). Improved temperature control systems, which ensure more precise regulation, reducing temperature overshoot and preventing short cycling and oscillations in large production freeze-dryers, have also been developed (Tenedini and Sutherland 1984).

The most advanced control system at the moment uses only the manipulation of shelf temperature, estimates the product temperature by an advanced PRT method, and uses the *LyoDriver* algorithm to calculate for a given prediction horizon the optimal sequence of shelf temperature values that will minimize the difference between the actual product temperature and its set-point value (Barresi et al. 2009a, 2009b).

Manipulation of shelf temperature is characterized by a slow response, as said before, caused by the large inertia of the shelf, while manipulation of the chamber pressure ensures fast response; in this way, it is possible to modify effectively the heat transfer coefficient. In fact, gas conduction varies linearly with the chamber pressure and represents the main contribution to heat transfer in the low-pressure conditions typical of primary drying (Pikal et al. 1984; Pisano et al. 2011c).

Neumann (1963) proposed to throttle the valve placed in the spool connecting the chamber to the condenser (or acting on condenser temperature) in order to maintain the product at the desired temperature.

Rieutord (1962) realized the controlled bleeding system, using an inert gas to control the pressure in the chamber; this is the system currently used to regulate chamber pressure, especially in industrial units (Nail and Gatlin 1985), and can be combined with end-of-sublimation determination (Chase 1998). If a mass flowmeter is added to the bleeding system, its output can be used to monitor the sublimation flux (Pisano et al. 2016).

Rieutord proposed a control loop feeding gas proportionally to the variation in electrical conductivity of the product; the shelf temperature could be kept constant or regulated independently, with independent temperature control of the different shelves.

Successively, Willemer (1987) showed an example of nominal-actual value regulation of shelf temperature, product temperature measurement by BTM, and its control by changes in chamber pressure, showing that pressure regulation in the chamber using "product gas" or nitrogen gas is equivalent from the point of view of drying time optimization.

Jefferis (1981) evidenced that the Rieutord's approach does not inherently balance the heat transfer rates (heat input from the shelf and heat removal by sublimation) and can become unstable during condenser saturation; a cascade control algorithm, in which dryer pressure was a function of product resistivity and condenser temperature, was thus elaborated. The algorithm was very stable in the first part of the cycle but failed to adequately balance the heat transfer rate in the last part of the cycle; it was also very complex to implement. A simpler algorithm, which ensured improved control, was proposed later on (Jefferis 1983).

Tenedini and Bart (1999) claimed that turning the pump on/off allowed energy to be saved, produced a higher level of product purity than the inert gas bleed system, and sped up freeze-drying as a consequence of the higher specific heat of the water vapor with respect to the inert gas (but it must be considered that during primary drying the water vapor is prevailing also in the case of inert bleed control). As a matter of fact, this method is nowadays generally adopted only in cheap laboratories and pilot scale units.

6.2.2 CONTROL LOGICS

The number of research papers concerning control logic in freeze-drying is relatively limited.

Meo and Friedly (1973) carried out an experimental study of the optimal feedback control of a freeze-dryer, obtaining a reduction of the drying time, but without discussing the control approach. Nail and Gatlin (1985) discussed the feedback control of primary drying using product temperature or electrical conductivity as measured variables.

The thermodynamic lyophilization control described by Oetjen (2004) is simply an automatic procedure that maintains the shelf temperature at a set value until the product temperature, measured with the BTM method, is below the required value, considering a safety margin; it prevents failure but does not allow optimization.

The Pikal's group (Tang et al. 2005) proposed and patented an expert system, named SMART™ Freeze-Dryer, for manipulating the shelf temperature, using the feedback information on product temperature obtained by means of the MTM algorithm (a simple PRT method) and some empirical and good practice rules; the chamber pressure was set at an "optimal value" calculated from a relationship as a function of the initial product temperature (measured) and kept constant.

An ideal model-based control strategy, where the shelf temperature is continuously changed in order to maintain the product temperature at a safe level, was investigated by Fissore et al. (2008), comparing the performances with that of an optimal constant pressure and shelf temperature policy and of a PI controller. The influence of the chamber pressure was investigated in both cases, showing that a value that minimizes the drying time may exist. The possibility of manipulating the pressure in the chamber to control quickly the maximum product temperature in case the control system malfunctioned was also studied.

6.2.2.1 Open-Loop Control

Many published works are about model-based open-loop control systems. The optimal cycle parameters are defined off-line by modeling and computational tools; a brief summary of the different models utilized is reported by Daraoui et al. (2010).

Liapis and Litchfield (1979) adopted a quasisteady state model and proposed to manipulate the radiator energy output and the total pressure in the drying chamber, giving as constraints the scorch temperature of the dried product and the melting temperature of the frozen interface. The approach was successively extended to the dynamic model case (Litchfield and Liapis 1982). The objective function was the minimization of the time required to get a fixed amount of residual water in the product at the end of the primary drying step. The control policy consisted in dividing the cycle into four segments, where each of the two manipulated variables were constant and tuned off-line. In fact, a simplified approach is taken, as in each segment the method evaluates if the process dynamics are controlled by the heat transfer from the shelf or by the mass transfer: in the first case the shelf temperature and in the second case the chamber pressure is manipulated.

In a similar approach, Lombraña and Díaz (1987a, 1987b) adopted a quasisteady partial differential equation model, using an experimentally derived mass-transfer relationship. A single manipulated variable (shelf temperature) and the multivariable case were investigated, showing that in the second case a reduction in the drying time can be obtained. Also in this case the temperature profile in the vial is assumed to be known and the algorithm determines whether the process is under mass or heat transfer control, thus activating the most suitable controller. Lombraña et al. (1997) adopted a finite element model to freeze-drying special foods in vials, to evaluate various heating strategies in a heat-controlled operating regime, for conditions in which the heat transfer mechanism was mainly radiant.

Lopez-Quiroga et al. (2012) applied time-scale analysis to a detailed first-principle-based model, proposing a simplified model able to describe the process at the time scales relevant to quality, which reduces the computational effort to calculate the optimal operational policy. Two different scenarios were considered, with only shelf temperature or both temperature and pressure manipulated.

Variational calculus and a detailed monodimensional model of the process have been used by Sadikoglu et al. (1998) to optimize both the primary and the secondary drying stage of freeze-drying of skim milk (used as a model of complex protein solution) in bulk. The optimal operational policy, varying both pressure and shelf temperature, was determined considering melt and scorch temperature as constraints. They found that in primary drying the chamber pressure must be maintained at a minimum, as the process was under mass transfer control; thus, the results obtained must be considered specific for the product and the conditions considered, while primary drying for most pharmaceutical products is actually carried out under heat transfer control. The work was extended to freeze-drying in vials, using a bidimensional detailed model; two different freeze-dryer designs were considered, a traditional one and a second one with independent control of heating plate temperatures (Sadikoglu et al. 2003; Sadikoglu 2005). Gan et al. (2004, 2005) considered the effect of the heterogeneity of the batch on the optimal heating policy, indicating minimum number of vials that have to be monitored by sensors and their relative location on the tray.

Finally, some other optimization approaches can be mentioned. Boss et al. (2004) used successive quadratic programming to carry out an open-loop model-based optimization for skim milk and soluble coffee; different constant optimal shelf temperatures and pressures were obtained for primary and secondary steps.

Kuu and Nail (2009) developed a computer program to determine the optimal shelf temperature and chamber pressure that minimize primary drying time, requiring that the product temperature profile be below a target value, and further optimized the cycle by dividing the drying time in a series of ramping steps for the shelf temperature in a cascading manner.

A model-based control algorithm was proposed by Trelea et al. (2007) to maximize process productivity and ensure product quality preservation. A user-friendly software was generated for the interactive selection of the operating conditions in primary and secondary drying, which takes into account the variation of the glass transition temperature with the residual water content.

A method and apparatus for optimizing primary drying, based on open-loop control after generating a designed primary cycle, was patented by Tchessalov and Warne (2007).

Open-loop optimization assumes that all the parameters and all the variables of the process can be perfectly known, and the model adopted describes perfectly the dynamics of the process. Unfortunately, this is rarely completely true, while many parameters may be the cause of deviations between the real case and the model: changes in the heating and cooling rate of the apparatus, in the batch load, or in the freezing conditions (and consequently in the cake structure of the product, which influences the resistance to mass flow) and radiation effects that can cause heterogeneity of the batch.

In the next sections model-based control logics with predictive capabilities and feedback corrections will be presented and discussed for each of the three steps of the freeze-drying process.

6.3 CONTROL OF FREEZING AND CONTROLLED NUCLEATION

The freezing step is crucial for the process and is also the most difficult step to be controlled because ice nucleation is a stochastic process: it has an energy barrier, so it does not occur spontaneously at the thermodynamic freezing temperature, but at a significantly lower temperature. Furthermore, the presence of impurity nuclei limits the degree of supercooling, which is usually different from vial to vial (and different in lab and production units). During freezing temperature gradients forms naturally in the liquid product, and as the formation of the ice crystals is exothermic, strong temperature changes can be observed with time (see Figure 6.1).

As it was shown that final crystal size strongly depends on the nucleation temperature, and thus on supercooling, and is influenced by freezing rate and temperature gradients, it is evident that it is essential to control this step to reduce inter- and intra-batch variability and to optimize the drying step. In fact, the size of the ice crystals determines the size of the pores, its distribution within the product, and finally, the cake resistance to mass flow, thus influencing the drying time and determining the product temperature. Smaller pores increase the cake resistance, thus prolonging primary drying and increasing the interface surface, which may be beneficial for water desorption in secondary drying but in some cases may impair the stability of

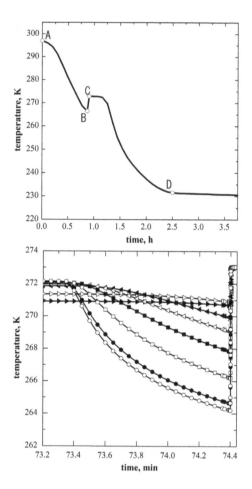

FIGURE 6.1   Example of modeling predictions of the product temperature during freezing of an aqueous solution of sucrose (5% by weight) in the case of (upper graph) shelf-ramped freezing and (lower graph) vacuum induces surface freezing (VISF). Three events are observable: (AB) supercooling, (BC) nucleation, and (CD) crystal growth. In the case of VISF, the decrease in pressure promotes the decrease in product temperature; the temperature profiles as predicted by the model at different vertical positions when pressure is decreased are shown: (O) $z = 0$ mm; (●) $z = 1$ mm; (□) $z = 2$ mm; (■) $z = 3$ mm; (◁) $z = 4$ mm; (◀) $z = 5$ mm; (▷) $z = 6$ mm, and (▶) $z = 8$ mm. [Reprinted from Pisano and Capozzi (2017), with permission from Elsevier.]

the active principle. The freezing step can also influence the development of different polymorphic forms of the crystallizing drug or excipients.

Unfortunately, there is generally a lack of control of the degree of supercooling, but it is evident that there is a strong interest in controlling this step, to improve product quality, and batch reproducibility and uniformity, and there is also a big potential for process optimization. In fact, depending on the characteristics of the product, the optimal pore size that reduces the total process time can be estimated.

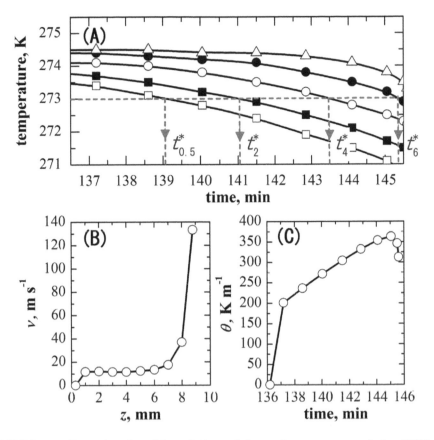

FIGURE 6.2 Examples of model predictions of the product temperature during VISF of 5% sucrose at different product depth: ($\square$) $z = 0.5$ mm; ($\blacksquare$) $z = 2$ mm; (O) $z = 4$ mm; ($\bullet$) $z = 6$ mm; and ($\triangle$) $z = 8$ mm. The plot shows also the time at which the product reaches its equilibrium freezing temperature, t_z^*. From this time, it is possible to determine the position of the freezing front over time and thus its rate and the freezing front rate (B). The product temperature gradient is also shown (C). Data refer to the freezing of mannitol 5% nucleated at 268 K [Reprinted from Pisano and Capozzi (2017), with permission from Elsevier.]

This motivated the research of these last years in freezing control. The fundamentals of freezing and the different techniques for controlling nucleation have been recently reviewed elsewhere (Pisano 2019), while the possible tools to tune, measure, and predict the impact of freezing on product morphology were discussed by Pisano et al. (2019).

A first-principle model for the freezing step can be very useful to predict different morphologies varying the operating conditions and, finally, to find the optimal conditions and the design space (Pisano and Capozzi 2017; Arsiccio et al. 2017; Arsiccio and Pisano 2018). Figures 6.1 and 6.2 shows how a predictive model for freezing can be used to estimate temperature profiles during vacuum and nucleation, temperature gradients, and freezing front velocity.

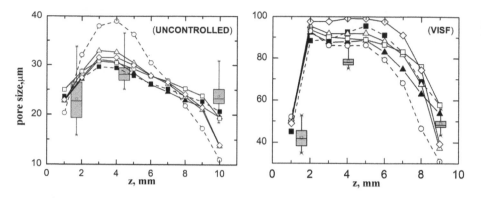

FIGURE 6.3 Comparison between detailed model predictions (solid line) and simplified model predictions (dashed line) by Arsiccio et al. and empirical laws proposed by Bomben and King ($\triangle$), Kochs et al. ($\square$), Kurz and Fischer ($\circ$, $\diamond$) in case of conventional and controlled (VISF) nucleation (mannitol 5%). SEM observations are reported in the box plots. [Reprinted with permission from Arsiccio et al. (2017). Copyright 2017 American Chemical Society.]

Figure 6.3 compares the experimental pore size distribution obtained using conventional and controlled freezing, showing the values predicted by different models.

A simple model of the freezing stage coupled with product temperature measurement was also used to roughly estimate the cake resistance, thus strongly improving the robustness of an observer used for primary drying monitoring (Bosca et al. 2015).

It must be said that it is not possible to control nucleation in a strict sense, but the term "controlled nucleation" is adopted to indicate a procedure that tries to control the freezing rate, in order to obtain a uniform temperature distribution in the vials of the batch and inside the single vials, and to induce the simultaneous nucleation at the desired level of supercooling in the whole batch. Both these actions are not an easy task, but recently a monitoring and control system based on the heat flux measuring tool (AccuFlux™) has been proposed to control the freezing rate, when the temperature is almost constant (Thompson 2013). Using different sensors placed on the shelf is also possible to consider, for the control action, the estimated temperature of vials in different positions. It has also been shown that suspended freezing can improve batch uniformity (Capozzi and Pisano 2018).

In Chapter 1 the main techniques proposed to induce nucleation at the desired temperature have been already presented and will be briefly commented here to evidence general limits and characteristics (a detailed comparison of the different patents and technical solutions proposed and commercially available can be found in Pisano et al. 2019).

Vacuum-induced surface freezing, depressurization methods, and ice-fog techniques trigger nucleation at the upper surface, inducing a directional solidification; all these techniques can be found implemented in commercially available equipment, but all suffer from some limitations.

The ice-fog technique initially had problems with sterilization of the stream used to generate the ice fog; this problem was overcome, but it may be difficult to distribute the fog uniformly, especially when the size of the apparatus increases, and the simultaneous nucleation of the whole batch is not ensured.

Vacuum-induced surface freezing is a free technology (while the other two have been patented) that can be easily retrofitted with laboratory equipment as it does not require overpressure or hardware modifications. It was initially abandoned, because it easily caused esthetic problems with the product, but it was shown that with the proper tuning of the operating conditions, which must be found for each product, this process can work efficiently, reducing the variability of the batch (Oddone et al. 2014, 2016; Arsiccio et al. 2018). The impact on primary and secondary drying duration was also investigated, showing that a significant reduction of the total drying time can be obtained (Oddone et al. 2017).

The technological limit of the methods based on fast pressure reduction is that the rate of depressurization is limited by the size of the chamber and that of the valve: currently both vacuum-induced freezing and depressurization work very nicely in small apparatus, but performances may decrease in larger ones.

Ultrasounds promote a global supercooling, acting on the bulk product; it is a very interesting technique, especially suitable to the lab, but very difficult to implement in industrial units.

6.4   PROCESS MONITORING AND PARAMETER ESTIMATION

Optimal control and optimization of freeze-drying is possible only if the process variables are continuously monitored (product temperature, desorption and sublimation rate, residual ice fraction, or moisture are generally measured or inferred) and if the values of the parameters characterizing the system dynamics (heat and mass transfer coefficients) are estimated in-line.

Since the early systems, the temperature was measured with temperature probes inserted in the tray or in some vials, but as discussed before, this method has strong limitations and gives information of limited utility and reliability. Alternatively, the batch average temperature can be estimated using the pressure rise test (PRT). No additional hardware is required, because a capacitive pressure transducer is needed, which is generally always available in pilot scale and production units, with a valve isolating the chamber from the external condenser (better if fast closing); this makes it easily retrofittable to existing freeze-dryers (Galan 2010).

In the early version, the ice temperature was calculated from the pressure measured at the end of the shutting-off period, with the vapor pressure curve of ice; the valve in the duct connecting the chamber with the condenser was closed, leaving the pressure to rise as a consequence of sublimation, until equilibrium with ice was reached (Oetjen et al. 1962). This method, called the barometric temperature measurement (BTM), was invasive and not very accurate; in fact, prolonging the test until equilibrium is reached causes a significant ice temperature rise, because the heat transfer coefficient increases with the chamber pressure.

A procedure that allowed reduction of the shut-off time and, thus, the temperature increase of ice, with a sufficiently precise temperature determination, was later on proposed by Oetjen et al. (1998); the PRT was interrupted when the first derivative of the pressure curve reached its maximum. Advanced versions of the pressure rise test, like the DPE or DPE+, which take advantage of a model of the process, not only give a more robust estimation of the batch temperature but also allow to obtain a regular estimation of the process parameters, namely the heat transfer coefficient and cake resistance (Velardi et al. 2008; Fissore et al. 2011a, 2018; Pisano et al. 2011b, 2017). Therefore, they can be easily used in a closed-loop control algorithm, like the already mentioned *LyoDriver*.

This controller is anyway compatible with other predictive tools, based on the process response principle. Examples include the pressure decrease test (Pisano et al. 2014) or the valveless monitoring system (Pisano et al. 2016): they are less dangerous for the product because no temperature rise occurs, and they require no moving parts; thus, they are ideal for large units with automatic loading and unloading systems, even if they are less powerful than the previous ones because they estimate directly the sublimation flux and recover the information on the product temperature only if the heat transfer coefficient is known.

The TDLAS installed in the duct (Gieseler et al. 2007a) is another PAT tool developed to measure the sublimation flow, and to recover the average batch temperature provided that the heat transfer coefficient is given (Schneid et al. 2009). It is mainly used for monitoring, as a control logic with the goal of maintaining a predetermined sublimation rate would be risky: in fact, a change in cake structure or in the heat transfer coefficient would cause an increase in the product temperature.

The heat flux measurement has been recently developed and applied to process monitoring and control: at steady state, the sublimation flow is proportional to the heat flow to vials or trays, and as in previous cases the product temperature can be estimated from preliminary determination of the heat transfer coefficient (Vollrath et al. 2017) [see also Chapter 7]. The advantage of this method is that it can monitor several small clusters of (or single) vials, thus allowing distributed control, or the whole batch, depending on the surface covered by the sensor, and the heat flux measurement is not dependent on loading, unlike the mass flow measured with the TDLAS. However, at the moment the limit of the heat flux sensor is that it does not detect all heat from radiation; thus, calibration may be necessary, and this may be dependent on apparatus and scale. To improve the performance of the heat flux sensor it was proposed to modify the bottom of the vials (Chen et al. 2008), or to insert between the sensor and the vials a metallic foil.

When the average temperature of the batch is estimated, it must be remembered that the mean value is generally close to that of the largest part of vials in the center of the shelf, but vials along the border may experience higher temperatures because of radiation from walls. To avoid the problem and the temperature gradients, some systems were proposed and patented, in which the apparatus walls are heated and cooled in synchronization with the interior of the apparatus (Connor 1995), or plates are inserted, connected to a second heat transfer circuit, and thermally isolated from the chamber wall (Oetjen and Haseley 2004; Sennhenn et al. 2003). These systems are very complicated and, to some extent, they may affect vapor-fluid dynamics; as a matter of fact, they have found scarce application.

Thus, when using the pressure rise-test–based methods, or one of the other monitoring tools that estimate an effective average temperature value of the batch, additional safety margins (which may be dependent on the size of the batch) must be introduced in the control algorithm, to be sure that product temperature remains below the limit value in the whole batch. To consider the temperature variance, it is also possible to improve the DPE algorithm simulating the batch as the sum of vials placed in the core and on the side of the batch, introducing a new parameter that is a complex nonlinear function of the variance (and of the covariance) of the other parameters (Barresi et al. 2010). Another possibility is to develop a tool that couples the results of the fluid dynamics of the water vapor inside the drying chamber obtained through computational fluid dynamics simulations and a detailed one-dimensional model of the drying, as discussed in Rasetto et al. (2010); by this way it would be possible to estimate the batch variance for certain operating conditions and tune the parameters of the controller for the best control strategy.

It is possible to evaluate in-line the values of model parameters also using a soft-sensor: it exploits the measurement of product temperature in a well-defined position, such as at the bottom of the vial, and a mathematical model of the process. The difference between the measured and the calculated values of product temperature is used to "correct" model equations in such a way that the difference is driven to zero. Both the high gain algorithm (Velardi et al. 2010) and the extended Kalman filter (Velardi et al. 2009; Bosca and Fissore 2011; Bosca et al. 2014) were proposed and tested, showing that the system is able to estimate the residual amount of ice in the vial, the resistance of the dried product to vapor flux (Bosca et al. 2013b), and the heat transfer coefficient from the shelf to the product in the vial (Bosca et al. 2013c). When comparing the soft sensor to the pressure-rise test, it has to be remarked that in the pressure-rise test–based algorithm, the parameters are estimated by solving a least-squares problem, which may be ill-conditioned, thus resulting, in certain cases, in inaccurate estimates (Fissore et al. 2011a); moreover, the response time may be an important issue, in particular for fully loaded batches (Pisano et al. 2017). With respect to the soft sensor, the main issue is related to the availability of a (sufficiently) accurate initial estimate of the monitored variables, and a method to overcome this issue was proposed by Bosca et al. (2015).

6.5   IN-LINE OPTIMIZATION

This section will focus on recently proposed advanced control approaches: *LyoDriver*, model predictive control (MPC), and soft sensors. They all have predictive capability and use mathematical modeling to account for the real process dynamics. Both *LyoDriver* (Pisano et al. 2010) and MPC (Pisano et al. 2011a) can be implemented using whichever sensor can supply the state and parameter values of the system; up to now they have been implemented coupled with the dynamic parameters estimation (DPE) algorithm, previously described.

Only the maximum allowable product temperature and cooling/heating characteristics of the apparatus must be known, while all the process parameters are estimated online every time a new control action is calculated: the deviation from desired and actual product temperature is considered, minimizing the temperature overshoot.

Figure 6.4 shows the sequence of steps (LHS) and the operating principle (RHS) of *LyoDriver*, which manipulates only the shelf temperature, with pressure set at a constant prefixed value. After each pressure rise test, the curve is fitted, the product bottom (T_B) and interface temperature (T_i) are estimated, and a new set-point temperature sequence for the shelf is calculated.

LyoDriver can apply two different control laws: the conventional proportional-integrative-derivative law and the model-based control law. The product temperature set point used for the calculation of the corrective action also accounts for the temperature increase as observed during a PRT, the residual predicted overshoot, and a safety margin specified by the user. If the model-based control logic is used, the algorithm is faster because the solution of an optimization problem is not required.

In Figures 6.5–6.7 examples of control action sequences calculated by *LyoDriver* using either the PID or the model-based feedback control law are shown. In particular Figure 6.5 shows successive sequences of control actions calculated by *LyoDriver* using a wrong value of the heat transfer coefficient; it can be noted that even a wrong

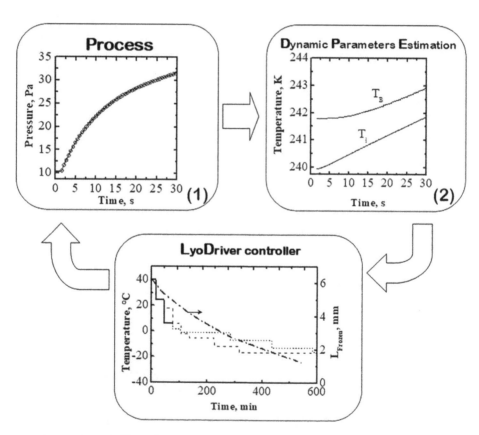

FIGURE 6.4 Sequence of steps (upper graph) and operating principle (lower graph) of the *LyoDriver* software coupled with DPE as state estimator. In the above plot the complete shelf temperature sequences predicted after the first (dashed line) and the second (dotted line) DPE are shown; the predicted evolution of the position interface is also shown [Reprinted from Barresi et al. (2009a) and Pisano et al. (2010), with permission from Elsevier.]

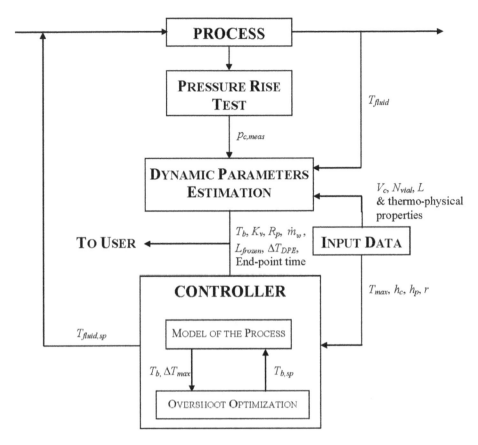

FIGURE 6.4 (Continued) Sequence of steps (upper graph) and operating principle (lower graph) of the *LyoDriver* software coupled with DPE as state estimator. In the above plot the complete shelf temperature sequences predicted after the first (dashed line) and the second (dotted line) DPE are shown; the predicted evolution of the position interface is also shown [Reprinted from Barresi et al. (2009a) and Pisano et al. (2010), with permission from Elsevier.]

estimation of a parameter does not prevent the control system from reaching the goal of preserving the product quality: in fact, after each control interval the state of the system is updated, and the controller can correct the control strategies, taking into account the difference between the real and the calculated status of the product. It can be noted that each time the control calculates a sequence of future control actions, based on the previsions made using the model and the actual parameter estimations, this calculation improves the quality of the control, but only the first action is actually implemented each time, because the whole sequence is recalculated at each step.

Figure 6.6 shows how the optimal shelf temperature sequence changes with the formulation for an enzyme, while Figure 6.7 (LHS) highlights the modification of

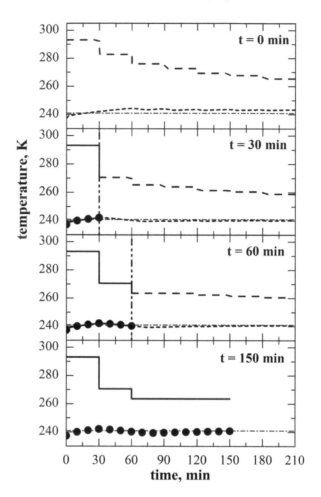

FIGURE 6.5 *LyoDriver:* Sequence of control actions calculated in successive steps when a PID feedback controller is used (T_{max} = 241 K, control interval = 30 min, prediction horizon equals the drying time); a wrong value of the heat transfer coefficient is used in this case, to demonstrate that the controller can work effectively anyway, because the system state is updated at the end of each control horizon. The sequence of fluid temperatures calculated by *LyoDriver* after each control horizon (long-dashed line) and the values applied (solid line) are shown. The product temperature at the bottom of the vial predicted by the model used by *LyoDriver* (short dashed line) when the value of K_v is affected by an error is also reported and compared with the correct evolution estimated through the same model (●) but using the correct value of K_v (15 Wm^{-2}K^{-1}) [Reprinted from Pisano et al. (2010), with permission from Elsevier.]

the cycle (with significant reduction of the drying time) passing from uncontrolled freezing to controlled nucleation.

It can be noted that in all cases shelf temperature rises at its maximum rate at the beginning of the drying (the maximum allowed fluid temperature can be set by the

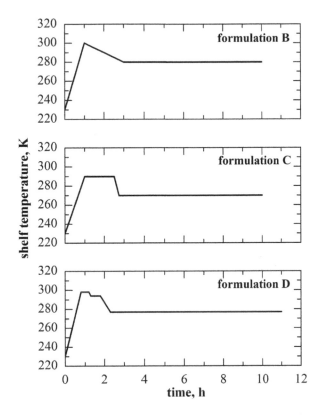

FIGURE 6.6 Examples of cycles automatically developed by means of *LyoDriver*, with a PID feedback controller, for the same enzyme with different excipient formulations: B, mannitol 4 %, sucrose 1%; C, lactose 1.25%, mannitol 2.5%, sucrose 1.25%; D, PVP 5% (collapse temperature determined by freeze-drying microscope is 263 K for B and C, 254 K for D) [Reprinted from Pisano et al. (2013b), with permission from Elsevier.]

user); here the product is far from its limit temperature; there is no risk of exceeding the limit temperature, and the cycle can be optimized, reducing the time necessary to reach stationary conditions, considering the thermal inertia of the system. A progressive increase of the shelf temperature, which is typical of the cycle developed by many practitioners, is strongly inefficient because it slows down the cycle, setting a low temperature (for misunderstood safety reasons) when the product temperature is lowest, just after the freezing step. Shelf temperature is then reduced, when product temperature approaches the limit, and the cake resistance increases. In the last part of the cycle the monitoring based on PRT is less effective, but shelf temperature can be maintained constant.

A temperature measurement–based system (the soft sensor) can be used as an alternative to pressure-rise test–based systems; a comparison of the performances can be found in the article by Bosca et al. (2016). The scheme of the control system,

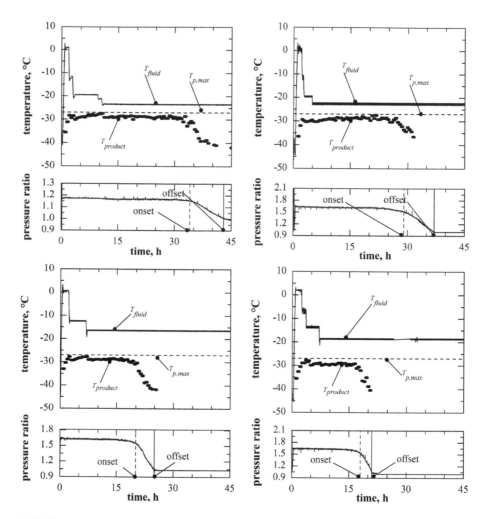

FIGURE 6.7 Freeze-drying cycles for mannitol-based solutions in the case of uncontrolled (top graphs) and controlled nucleation (bottom graphs). (LHS) Cycles controlled by *LyoDriver*, with model-based feedback controller, in the case of 10% mannitol solution. The evolution is shown for the temperature of the heat-transfer fluid (solid line), the product temperature (●), and for the pressure ratio between Pirani and Baratron, which allows estimation of the end of primary drying (onset-offset range).

(RHS) Simulation of the lyophilization of 5% mannitol solution using the process parameters obtained in-line by *LyoDriver*. The evolution is shown for the fluid temperature (solid line), T_B (■), and the thickness of the frozen layer. [Reprinted with permission from Oddone et al. (2014). Copyright 2014 American Chemical Society.]

with the option of distributed control also, is given in Figure 6.8, while Figure 6.9 shows an example of a cycle optimized in-line using the soft-sensor, and its comparison with the case of constant shelf temperature.

Bosca et al. (2013c) proposed a control system for the in-line optimization of the temperature of the heating shelf, using the soft-sensor. Briefly, once the

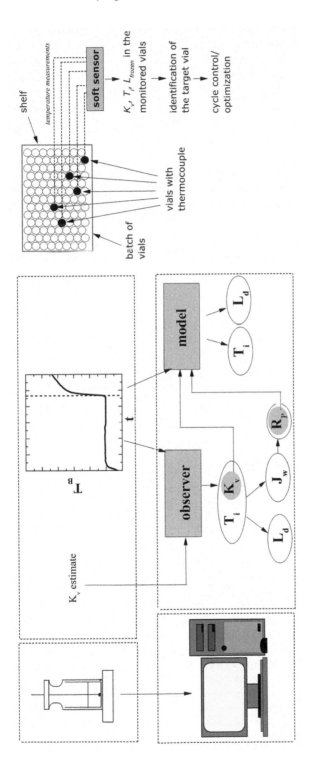

FIGURE 6.8 (LHS) Sketch of the working principle of the soft-sensor. [Reprinted from Bosca et al. (2013c), with permission from Elsevier.] (RHS) Sketch of the control system based on the use of the soft sensor to track product dynamics in various vials of the batch [Reprinted from Bosca et al. (2013a) by permission of the publisher (Taylor & Francis Ltd, www.tandfonline.com).]

values of K_v and R_p are estimated by the soft sensor, the algorithm calculates, for the selected pressure, the design space of the process, that is, the range of values of T_{shelf} that allows maintenance of the product temperature below the limit value ($T_{p,max}$). At the beginning of the primary drying stage, when the product is still completely frozen, the T_{shelf} is set to an arbitrarily high value, such as 0°C, and then, with every control internal, it is modified using the estimated values of K_v and R_p to calculate the design space and, from this, the limit shelf temperature to keep the product temperature as close as possible to the target value (Bosca et al. 2013c, 2016).

With multiple temperature-based soft sensors to monitor selected vials in different positions, it is possible to account for batch nonuniformity, selecting operating conditions for the most critical vials: in the first part of the primary dying, for those

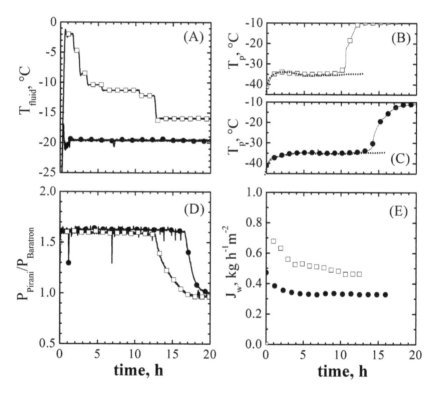

FIGURE 6.9   Comparison between the results obtained in a freeze-drying cycle (for a 5% by weight sucrose aqueous solution, $P_c = 10$ Pa) carried out with a constant value of the shelf temperature (●, $T_{fluid} = -20$°C) and with a shelf temperature optimized using the soft sensor (□). Graph A, values of shelf temperature vs. time; graphs B and C, product temperature detected with a thermocouple in close contact with vial bottom and values estimated using the soft sensor (dotted lines); graph D, ratio between pressure signals from Pirani and Baratron gauges; graph E, sublimation flux measured with the pressure rise test [Reprinted from Bosca et al. (2013c), with permission from Elsevier.]

where product temperature is higher (as a consequence of higher heating flux due to radiation or contact), and finally for the rest of the batch, typically the core, to complete drying in the shortest time. A system of this type has been described by Bosca et al. (2013a) and is shown in Figure 6.8 (RHS).

It is also possible to calculate a control action that considers the mean product temperature and the temperature of the radiated vials, coupling into a hybrid sytem the pressure rise method and a set of observers that estimates the batch variance (Barresi et al. 2009b).

The product temperature can be controlled effectively by manipulating the shelf temperature until drying is limited by heat transfer; when sublimation is controlled by mass transfer, T_{shelf} approaches product temperature and the sublimation rate is reduced to very low values. Such a situation can easily occur with high solid content or when the allowed maximum temperature of the product is very low. An example, for a cycle controlled by *LyoDriver*, is shown in Figure 6.10 (bottom left graphs); a pressure reduction is required, to restore a good sublimation rate, and when this takes place, the controller set the shelf temperature at higher values (Barresi et al. 2009a).

The control system discussed above manipulates only the shelf temperature, maintaining pressure at a constant prefixed value. It is evident that, in addition to avoiding mass transfer limitations, the optimal value of pressure might change during the drying step to minimize drying time. Thus, better optimization would require more sophisticated control strategies that can adjust both shelf temperature and pressure. An example was given by Pisano et al. (2011a) who designed a model predictive control (MPC) strategy that maintains the product temperature at its target value by adjusting both shelf temperature and pressure (see Figure 6.11). The corrective action is calculated based on the difference between the set-point value and the experimental measurement of the product temperature and can include any product or process constraints through appropriate penalty functions. For example, Pisano et al. (2011a) showed how the MPC can account for the maximum vapor flow rate that can be evacuated from the drying chamber into the calculation of the corrective action. Further constraints can be included, such as on the magnitude or number of changes in process conditions.

Other authors proposed the use of model predictive control strategy to the control of the lyophilization process, but they did not have any success in practice for various reasons.

The manipulation of pressure has a positive impact on the drying performance. For example, Figure 6.12 shows that, if the controller adjusts only the shelf temperature, the drying time remains unchanged whichever algorithm is used: PID as in *LyoDryver* or MPC. By contrast, the manipulation of both shelf temperature and pressure by MPC gives the highest rate of sublimation and hence the shortest drying time (Pisano et al. 2011a). However, the MPC manipulation of pressure can be implemented only if the pressure dependence of the heat transfer coefficient is known, which means that the equipment-vial system has been fully characterized. For example, PRT can estimate the heat transfer coefficient in real time, but that estimate refers to a given pressure while its pressure dependence remains unknown. In conclusion, the tuning of MPC is generally more complex than that of a simple

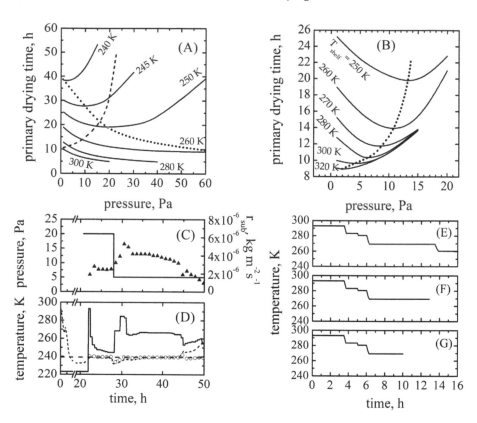

FIGURE 6.10 Upper graphs: Effect of the chamber pressure and of the heating shelf temperature on the primary drying time in the case of constant shelf temperature (graph A) and adopting the model-based control strategy that maintains the maximum product temperature at a value lower than 240 K (Graph B). The locus corresponding to the minimum of the primary drying time for the various shelf temperatures is also shown (dotted line). In graph A, the dashed line corresponds to the values of chamber pressure and of shelf temperature that satisfy the constraint on the maximum product temperature. Case study: BSA solution. [Reprinted from Fissore et al. (2008), with permission from Elsevier.]

Bottom left graphs: Example of the results obtained during an FD cycle run using *LyoDriver* to monitor and control PD stage (glass vials filled with a 10% by weight sucrose solution on tray). After freezing, the chamber pressure was set at 20 Pa and then lowered to 5 Pa after about 5 h. The fluid temperature is set by *LyoDriver* to maximize the sublimation rate; the maximum allowable product temperature has been set to 240 K, corresponding to T_g' of the freeze-dried solution. Graph C: Chamber pressure (solid line) and sublimation rate (▲) evolution estimated through PRT. Graph D: Comparison between temperature at vial bottom estimated by DPE (○) and measured through thermocouples (short-dashed line). The maximum product temperature allowable (equal to 240 K) and the set-point fluid temperature sequence calculated through *LyoDriver* (solid line) have been also reported [Reprinted from Barresi et al. (2009a), with permission from Elsevier.]

Bottom right graphs: Time evolution of T_{fluid} (solid line) when the chamber pressure is varied in a cycle controlled using the model-based controller ($T_{max} = 250$ K). The pressure at the beginning is 20 Pa. Graph E: P_c is not changed. Graph F: P_c becomes 15 Pa after 6 h. Graph G: P_c becomes 10 Pa after 6 h.

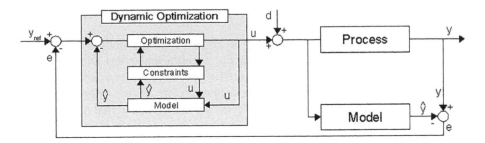

FIGURE 6.11   Block diagram of the model predictive control (internal model control) system proposed by Pisano et al. (2011a) [Reprinted with permission from Pisano et al. (2011a). Copyright 2011 American Chemical Society.]

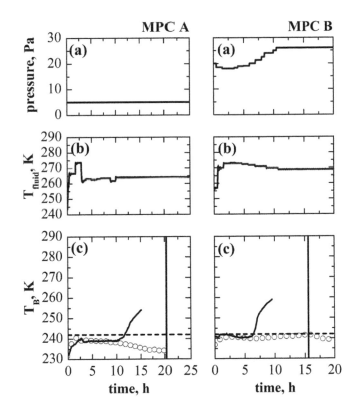

FIGURE 6.12   Cycles designed by (left-side graphs) MPC A (manipulation of shelf temperature) and (right-side graphs) MPC B (manipulation of shelf temperature and pressure) algorithms for the freeze-drying of a 5% (w/w) sucrose solution. Evolution of (graph a) chamber pressure, (graph b) fluid temperature, and (graph c) product temperature at the vial bottom as measured by thermocouples (solid line) and estimated by PRT technique (O). The horizontal dashed line in graph c indicates the maximum allowable product temperature, while the vertical line indicates the primary drying endpoint [Reprinted from Pisano et al. (2013a) by permission of the publisher (Taylor & Francis Ltd, www.tandfonline.com).]

proportional control law (*LyoDriver*) as widely discussed in the article by Pisano et al. (2011a).

Both control systems can effectively be implemented on both laboratory and industrial scale freeze-dryer, minimizing the experimental effort and the human resources on the plant. Furthermore, if it is true that both *LyoDriver* and MPC have predictive capacity, MPC is the only system that can effectively manage equipment constraints and compensate errors in model prediction by the internal model control strategy.

6.6 CYCLE DEVELOPMENT

Automatic control systems found up to now limited application in industrial units but have been commonly used at laboratory scale, where they can be used for process optimization and process intensification (Barresi and Pisano 2014). In fact, they allow both the determination in-line of base cycles and of reliable heat and mass transfer parameters, which can then be used for scale-up (Fissore and Barresi 2011) and cycle development using off-line approaches (Fissore et al. 2012a, 2015).

Various "control strategies" using the PRT outcomes have been proposed in the literature and can be classified into two classes. A first group includes the thermodynamic lyophilization control, first reported by Oetjen (1999), and the Smart™ Freeze Dryer (Tang et al. 2005; Gieseler et al. 2007b). In both cases, the control law is based on a set of heuristics, without any predictive capability. These algorithms can hardly be classified as feedback control systems, but more as expert systems, as they do not allow in-line control applications but may be suitable for cycle development.

The other group is constituted by the same closed-loop control logics described in the previous section. An apparatus equipped with an automatic closed-loop control, like *LyoDriver*, can be used to develop in a few steps (or even with a single run) a close-to-optimal cycle; only the maximum allowable product temperature must be specified, and constraints on the maximum shelf temperature and on the number and type of steps can be fixed.

Same of the examples shown before to illustrate the performances of the controller can also be taken as examples of cycle development. In Figure 6.6 cycles developed for the same enzyme but with different formulation are compared. The cycles were directly obtained by *LyoDriver*, using a PID feedback controller, and simplified by ignoring minor changes in the temperature of the heat transfer fluid. The differences in the sequence of shelf temperature set points reflect the different thermal characteristics of the excipients (mannitol, sucrose, lactose, and PVP) (Pisano et al. 2013b). Figure 6.7 (LHS) compares two cycles obtained for the same formulation (10% mannitol solution) with controlled and uncontrolled ice nucleation, evidencing how an optimized cycle can be obtained automatically when the permeability characteristics of the solid cake are modified, acting on the freezing step (Oddone et al. 2014).

If the industrial apparatus is also equipped with the monitoring and control tools described in the previous sections, obviously these can also be employed to adapt and transfer the cycle for it, overcoming the well-known scale up issues (Barresi 2011). It would be sufficient to launch a cycle imposing the proper restrictions on the product temperature.

The same tools can be adopted also to recover in-line the process parameters, to develop off-line optimization strategies or design spaces; safety margins can be handled in both cases, but in a different way, as discussed in the studies by Fissore et al. (2012a) and Pisano et al. (2013a), where off-line and in-line approaches have been compared.

It must be remembered that the design space can also be obtained in-line using a soft sensor (Bosca et al. 2013c). It has been shown how the design space approach can be used to handle the effect of the freezing protocol and of batch nonuniformity (Pisano et al. 2013c).

Alternatively, the same parameters obtained can be used to simulate drying behavior with a mathematical model of the process; Figure 6.7 (RHS) shows an example for the various freezing protocols (Oddone et al. 2014).

All the methods previously described find the optimal shelf temperature sequence for a given pressure, which can be identified by an off-line optimization procedure (Fissore et al. 2008, 2009a). Obviously, the result depends on the control logic adopted, but the system is not very sensitive (the range of pressure to be considered is also generally quite limited), provided the shelf temperature is varied. Figure 6.10 (upper graphs) shows an example, comparing the constant shelf policy and an ideal controller (Fissore et al. 2008); the product temperature constraint is implicitly considered in the second case, while the region where this limit is not trespassed must be delimited in the first case (see the dashed line). Other examples for different feedback and model-based controllers are reported by Fissore et al. (2009a).

If *LyoDriver* is used for cycle development, pressure should be manually adjusted by the user as soon as there is evidence to suggest that sublimation is rate controlled by mass transfer; the new optimal value can be calculated as discussed above, and an automatic switch can also be programmed. Figure 6.10 (bottom graphs) shows examples of a pressure switch, evidencing how a significant reduction of the drying time can be obtained with a proper selection of different pressures during primary drying. However, this strategy requires that the pressure dependence of the heat transfer coefficient is known a priori, but unfortunately these data are not always available and are both equipment and container specific.

6.7 SECONDARY DRYING OPTIMIZATION AND CONTROL

In secondary drying the residual moisture is generally the monitored variable (sometimes also the scorch temperature is considered), while the shelf temperature is generally the only manipulated variable even if often it is kept constant (at a higher set temperature than in primary drying). The chamber pressure has a week influence in secondary drying, but water partial pressure can influence the residual moisture in the product (Searles et al. 2017). As the process is usually heat transfer controlled in this stage, it is not very useful to keep pressure at a very low value; Sadikoglu et al. (1998) and Sadikoglu (2005) proposed to vary both shelf temperature and chamber pressure, increasing pressure to improve heat transfer and reduce drying time.

The residual moisture of the batch can be estimated from the composition of the chamber gas or from the desorption rate; the different proposed techniques have been discussed in Barresi et al. (2018), but the main limit is that they are not general and a specific calibration must be carried out for each product. Methods based on the pressure rise test have been developed also for secondary drying. Oetjen (1997) proposed to use two successive measurements of desorption rate to extrapolate the time at which the desorption rate reaches a given small value, assuming implicitly that the desorption rate is proportional to the residual moisture and that the value of the desorption kinetic constant does not change throughout the secondary drying. The method is very simple and can be used for a first estimation, but as a consequence of the very simplified approach, it fails in the last part of secondary drying, because of the variation of product temperature and of the surface area of the pores, owing to the heterogeneity of the freeze-dried solid.

An innovative approach has been patented by Fissore et al. (2011b): the key feature of the method is the coupling of the measurement of the desorption rate, obtained by means of the PRT or other devices, with a mathematical model of the process, thus obtaining a soft sensor that estimates in-line the desorption constant and the residual amount of solvent in the product at the end of primary drying. Figure 6.13 (LHS) shows a comparison of the predictions obtained with the two methods of the time required to complete secondary drying; it can be seen that the second one quickly converges to the exact value.

The parameters estimated can be used to obtain the design space also for secondary drying, but the soft sensor can also be integrated inside a control loop that

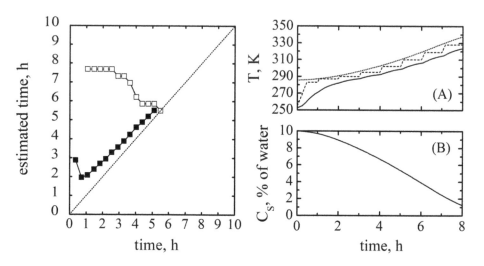

FIGURE 6.13 LHS: Comparison between the time required to complete secondary drying calculated using the method of Oetjen (■) and using the method proposed by Fissore et al. (□) for a small batch of vials containing a sucrose-based solution. Stop criterion: residual moisture concentration lower than 0.2% [Reprinted from Fissore et al. (2011b), with permission from Elsevier]. RHS: Example of cycle optimization for the secondary drying step. Graph A: evolution of the set point of the fluid temperature (dashed line), of product temperature (solid line), and of the limit product temperature (dotted line) *vs.* time. Graph B: evolution of residual moisture *vs.* time [Reprinted from Pisano et al. (2012) by permission of the publisher (Taylor & Francis Ltd, www.tandfonline.com).]

determines the optimal heating strategy for the secondary drying step (Pisano et al. 2012). In fact, the maximum allowed product temperature of the lyophilized product increases with a decrease in the residual moisture; if this relationship is known, the cycle can be optimized, increasing progressively the shelf temperature as long as secondary drying goes on and the residual moisture decreases. Figure 6.13 (RHS) shows an example of cycle development, evidencing that the solid temperature remains always below the collapse temperature.

ACKNOWLEDGMENTS

The contribution of Salvatore Velardi, Serena Bosca, and Irene Oddone, from Politecnico di Torino, is greatly acknowledged.

REFERENCES

Arsiccio, A., and R. Pisano. 2018. Application of the Quality by Design approach to the freezing step of freeze-drying: Building the design space. *J. Pharm. Sci.* 107:1586–1596.

Arsiccio, A., Barresi, A. A., and R. Pisano. 2017. Prediction of ice crystal size distribution after freezing of pharmaceutical solutions. *Crystal Growth Des.* 17:4573–4581.

Arsiccio, A., Barresi, A. A., De Beer, T., Oddone, I., Van Bockstal, P.-J., and R. Pisano. 2018. Vacuum Induced Surface Freezing as an effective method for improved inter- and intra-vial product homogeneity. *Eur. J. Pharm. Biopharm.* 128:210–219.

Barresi, A. 2011. Overcoming common scale-up issues. *Pharm. Technol. Eur.* 23:26–29. *Pharm. Technol. Eur. (PTE Digital)* (July 2011). www.pharmtech.com/ptedigital0711

Barresi, A. A., and D. Fissore. 2011. In-line product quality control of pharmaceuticals in freeze-drying processes. In *Modern drying technology*, Vol. 3: *Product quality and formulation*, ed. E. Tsotsas, and A. S. Mujumdar, 91–154. Weinheim: Wiley-VCH Verlag GmbH & Co. KGaA.

Barresi, A. A., and R. Pisano. 2014. Recent advances in process optimization of vacuum freeze-drying. *Am. Pharm. Rev.* 17:56–60.

Barresi, A. A., Pisano, R., and D. Fissore. 2018. Advanced control in freeze-drying. In *Intelligent control in drying*, ed. A. Martynenko, and A. Bück, Chap. 19, 367–401. Boca Raton, FL: CRC Press (Taylor & Francis Group).

Barresi, A. A., Pisano, R., Fissore, D., et al. 2009a. Monitoring of the primary drying of a lyophilization process in vials. *Chem. Eng. Proc.* 48:408–423.

Barresi, A. A., Pisano, R., Rasetto, V., Fissore, D., and D. L. Marchisio. 2010. Model-based monitoring and control of industrial freeze-drying processes: Effect of batch nonuniformity. *Drying Technol.* 28:577–590.

Barresi, A. A., Velardi, S. A., Pisano, R., Rasetto, V., Vallan, A., and M. Galan. 2009b. In-line control of the lyophilization process. A gentle PAT approach using software sensors. *Int. J. Refrig.* 32:1003–1014.

Bosca, S., and D. Fissore. 2011. Design and validation of an innovative soft-sensor for pharmaceuticals freeze-drying monitoring. *Chem. Eng. Sci.* 66:5127–5136.

Bosca, S., Barresi, A. A., and D. Fissore. 2013b. Use of a soft-sensor for the fast estimation of dried cake resistance during a freeze-drying cycle. *Int. J. Pharm.* 451:23–33.

Bosca, S., Barresi, A. A., and D. Fissore. 2013c. Fast freeze-drying cycle design and optimization using a PAT based on the measurement of product temperature. *Eur. J. Pharm. Biopharm.* 85:253–262.

Bosca, S., Barresi, A. A., and D. Fissore. 2014. Use of soft-sensors to monitor a pharmaceuticals freeze-drying process in vials. *Pharm. Dev. Tech.* 19:148–159.

Bosca, S., Barresi, A. A., and D. Fissore. 2015. Design of a robust soft-sensor to monitor in-line a freeze-drying process. *Drying Technol.* 33:1039–1050.

Bosca, S., Barresi, A. A., and D. Fissore. 2016. On the use of model-based tools to optimize in-line a pharmaceuticals freeze-drying process. *Drying Technol.* 34:1831–1842.

Bosca, S., Corbellini, S., Barresi, A. A., and D. Fissore 2013a. Freeze-drying monitoring using a new Process Analytical Technology: Toward a "zero defect" process. *Drying Technol.* 31:1744–1755.

Boss, E. A., Filho, R. M., and E. C. Vasco de Toledo. 2004. Freeze drying process: Real time model and optimization. *Chem. Eng. Process.* 43:1475–1485.

Capozzi, L. C., and R. Pisano. 2018. Looking inside the 'Black Box': Freezing engineering to ensure the quality of freeze-dried biopharmaceuticals. *Eur. J. Pharm. Biopharm.* 129:58–65.

Chase, D. R. 1998. Monitoring and control of a lyophilization process using a mass flow controller. *Pharm. Eng.*18:11–17.

Chen, R., Gatlin, L. A., Kramer, T., and E. Y. Shalaev. 2008. Comparative rates of freeze-drying for lactose and sucrose solutions as measured by photographic recording, product temperature, and heat flux transducer. *Pharm. Dev. Technol.* 13:367–374.

Connor, J. T. 1995. Process and apparatus for desiccation. U. S. Patent 5398426 A filed December 29, 1995.

Corbellini, S., Parvis, M., and A. Vallan. 2010. In-process temperature mapping system for industrial freeze-dryers. *IEEE Trans. Inst. Meas.* 59:1134–1140.

Daraoui, N., Dufour, P., Hammouri, H., and A. Hottot. 2010. Model predictive control during the primary drying stage of lyophilisation. *Control Eng. Pract.* 18:483–494.

De Beer, T., Burggraeve, A., Fonteyne, M., Saerens, L., Remon, J. P., and C. Vervaet. 2011. Near infrared and Raman spectroscopy for the in-process monitoring of pharmaceutical production processes. *Int. J. Pharm.* 417:32–47.

De Beer, T. R. M., Vercruysse, P., Burggraeve, A. et al. 2009. In-line and real-time process monitoring of a freeze-drying process using Raman and NIR spectroscopy as complementary process analytical technology (PAT) tools. *J. Pharm. Sci.* 98:3430–3446.

Fissore, D., and A. A. Barresi. 2011. Scale-up and process transfer of freeze-drying recipes. *Drying Technol.* 29:1673–1684.

Fissore, D., Velardi, S. A., and A. A. Barresi. 2008. In-line control of a freeze-drying process in vials. *Drying Technol.* 26:685–694.

Fissore, D., Pisano, R., and A. A. Barresi. 2009a, On the design of an in-line control system for a vial freeze-drying process: The role of chamber pressure. *Chem. Prod. Proc. Modeling* 4: Article 9, 21 pp.

Fissore, D., Pisano, R., Velardi, S. A., Barresi, A. A, and M. Galan. 2009b. PAT Tools for the optimization of the freeze-drying process. *Pharm. Eng.* 29:58–70.

Fissore, D., Pisano, R., and A. A. Barresi. 2011a. On the methods based on the pressure rise test for monitoring a freeze-drying process. *Drying Technol.* 29:73–90.

Fissore, D., Pisano, R., and A. A. Barresi. 2011b. Monitoring of the secondary drying in freeze-drying of pharmaceuticals. *J. Pharm. Sci.* 100:732–742.

Fissore, D., Pisano, R., and A. A. Barresi 2012a. A model based framework to optimize pharmaceuticals freeze-drying. *Drying Technol.* 30:946–958.

Fissore, D., Pisano, R., and A. A. Barresi. 2012b. A model-based framework for the analysis of failure consequences in a freeze-drying process. *Ind. Eng. Chem. Res.* 51:12386–12397.

Fissore, D., Pisano, R., and A. A. Barresi. 2015. Using mathematical modeling and prior knowledge for QbD in freeze-drying processes. In *Quality by design for biopharmaceutical drug product development*, ed. F. Jameel, S. Hershenson, M. A. Khan, and S. Martin-Moe, 565–593. New York: Springer.

Fissore, D., Pisano, R., and A. A. Barresi. 2018. Process Analytical Technology for monitoring pharmaceuticals freeze-drying – A comprehensive review. *Drying Technol.* 36:1839–1865.

Food and Drug Administration. 2004. Guidance for industry PAT – A framework for innovative pharmaceutical manufacturing and quality assurance. www.fda.gov/downloads/drugs/guidances/ucm070305.pdf (accessed May 2019).

Galan, M. 2010. Monitoring and control of industrial freeze-drying operations: The challenge of implementation Quality-by-Design (QbD). In *Freeze-drying/lyophilization of pharmaceuticals and biological products*, 3rd rev. Edition, ed. L. Rey, and J. C. May, Chap. 19, 441–459. New York: Informa Healthcare.

Gan, K. H., Bruttini, R., Crosser, O. K., and A. I. Liapis. 2004. Heating policies during the primary and secondary drying stages of the lyophilization process in vials: Effects of the arrangement of vials in clusters of square and hexagonal arrays on trays. *Drying Technol.* 22:1539–1575.

Gan, K. H., Bruttini, R., Crosser, O. K., and A. A. Liapis. 2005. Freeze-drying of pharmaceuticals in vials on trays: Effects of drying chamber wall temperature and tray side on lyophilization performance. *Int. J. Heat Mass Transfer* 48:1675–1687.

Gieseler, H., Kessler, W. J., Finson, M., et al. 2007a. Evaluation of tunable diode laser absorption spectroscopy for in-process water vapor mass flux measurements during freeze drying. *J. Pharm. Sci.* 96:1776–1793.

Gieseler, H., Kramer, T., and M. J. Pikal. 2007b. Use of Manometric Temperature Measurement (MTM) and SMART™ Freeze Dryer technology for development of an optimized freeze-drying cycle. *J. Pharm. Sci.* 96:3402–3418.

Grassini, S., Parvis, M., and A. A. Barresi. 2013. Inert thermocouple with nanometric thickness for lyophilization monitoring. *IEEE Trans. Inst. Meas.* 62:1276–1283.

Jefferis, R. P. III. 1981. Control of biochemical recovery processes. *Ann. NY Acad. Sci.* 369:275–284.

Jefferis, R. P. III. 1983. The microcomputer control of lyophilization. *Ann. NY Acad. Sci.* 413:283–289.

Jennings, T. A. 1999. *Lyophilization: Introduction and basic principles.* Boca Raton, FL: Interpharm/CRC Press.

Johnson, R. E., Gieseler, H., Teagarden, D. L., and L. M. Lewis. 2009. Analytical accessories for formulation and process development in freeze-drying. *Am. Pharm. Rev.* 12:54–60.

Kuu, W. Y., and S. L. Nail. 2009. Rapid freeze-drying cycle optimization using computer programs developed based on heat and mass transfer models and facilitated by Tunable Diode Laser Absorption Spectroscopy (TDLAS). *J. Pharm. Sci.* 98:3469–3482.

Liapis, A. I., and R. J. Litchfield. 1979. Optimal control of a freeze dryer – I. Theoretical development and quasi steady-state analysis. *Chem. Eng. Sci.* 34:975–981.

Litchfield, R. J., and A. I. Liapis. 1982. Optimal control of a freeze dryer – II. Dynamic analysis. *Chem. Eng. Sci.* 37:45–55.

Lombraña, J. I., and J. M. Díaz. 1987a. Heat programming to improve efficiency in a batch freeze-dryer. *Chem. Eng. J.* 35:B23–30.

Lombraña, J. I., and J. M. Díaz. 1987b. Coupled vacuum and heating power control for freeze-drying time reduction of solutions in phials. *Vacuum* 37:473–476.

Lombraña, J. I., De Elvira, C., and C. Villaran. 1997. Analysis of operating strategies in the production of special foods in vials by freeze drying. *Int. J. Food Sci. Technol.* 32:107–115.

Lopez-Quiroga, E., Antelo, L. T., and A. A. Alonso. 2012. Time-scale modeling and optimal control of freeze-drying. *J. Food Eng.* 111:655–666.

Meo, D. III and J. C. Friedly. 1973. An experimental study of the optimal feedback control of a freeze dryer. *J. Food Sci.* 38:826–830.

Nail, S., and L. A. Gatlin. 1985. Advances in control of production freeze dryers. *PDA J.* 39:16–27.

Nail, S., Tchessalov, S., Shalaev, E., et al. 2017. Recommended best practices for process monitoring instrumentation in pharmaceutical freeze drying – 2017. *AAPS PharmSciTech* 18:2379–2393.

Neumann, K. H. 1963. Temperature responsive freeze-drying method and apparatus. U. S. Patent 3077036 filed January 8, 1963.

Oddone, I., Barresi, A. A., and R. Pisano. 2017. Influence of controlled ice nucleation on the freeze-drying of pharmaceutical products: The secondary drying step. *Int. J. Pharm.* 524:134–140.

Oddone, I., Fulginiti, D., Barresi, A. A., Grassini, S., and R. Pisano. 2015. Non-invasive temperature monitoring in freeze drying: Control of freezing as a case study. *Drying Technol.* 33:1621–1630.

Oddone, I., Pisano, R., Bullich, R., and P. Stewart. 2014. Vacuum-Induced Nucleation as a method for freeze-drying cycle optimization. *Ind. Eng. Chem. Res.* 53:18236–18244.

Oddone, I., Van Bockstal, P.-J., De Beer, T., and R. Pisano. 2016. Impact of Vacuum-Induced Surface Freezing on inter- and intra-vial heterogeneity. *Eur. J. Pharm. Biopharm.* 103:167–178.

Oetjen, G. W. 1997. Method of determining residual moisture content during secondary drying in a freeze-drying process. European Patent EP0811153 B1 files December 10, 1997.

Oetjen, G. W. 1999. *Freeze-drying*. Weinheim: Wiely-VHC.

Oetjen, G. W. 2004. Industrial freeze-drying for pharmaceutical applications. In *Freeze-drying/Lyophilization of pharmaceuticals and biological products*, 2rd rev. Edition, ed. L. Rey, and J. C. May, Chap. 15, 425–476. Boca Raton, FL: CRC Press.

Oetjen, G. W., and P. Haseley. 2004. *Freeze-drying*, 2nd Edition. Weinheim: Wiely-VHC.

Oetjen, G. W., Ehlers, H., Hackenberg, U., Moll, J., and K. H. Neumann. 1962. Temperature-measurement and control of freeze-drying processes. In *Freeze-drying of foods*, ed. F. R. Fisher, 178–190. Washington, DC: National Academy of Sciences – National Research Council.

Oetjen, G. W., Haseley, P., Klutsch, H., and M. Leineweber. 1998. Method for controlling a freeze-drying process. U. S. Patent US6163979 A filed April 21, 1998.

Patel, S. M., and M. J. Pikal. 2009. Process Analytical Technologies (PAT) in freeze-drying of parenteral products. *Pharm. Dev. Technol.* 14:567–587.

Patel, S. M., Doen, T., and M. J. Pikal. 2010. Determination of the end point of primary drying in freeze-drying process control. *AAPS PharmSciTech* 11:73–84.

Pikal, M. J., Roy, M. L., and S. Shah. 1984. Mass and heat transfer in vial freeze-drying of pharmaceuticals: Role of the vial. *J. Pharm. Sci.* 73:1224–1237.

Pisano, R. 2019. Alternative methods of controlling nucleation in freeze-drying. In *Lyophilization of pharmaceuticals and biologicals: New technologies and approaches*, ed. K. R. Ward, and P. Matejtschuk, P., Chap. 4, 74–111. New York: Springer Science+Business Media.

Pisano, R., and L. Capozzi. 2017. Prediction of product morphology of lyophilized drugs in the case of Vacuum Induced Surface Freezing. *Chem. Eng. Res. Des.* 125:119–129.

Pisano, R., Arsiccio, A., Nakagawa, K., and A. A. Barresi. 2019. Tuning, measurement and prediction of the impact of freezing on product morphology: A step towards improved design of freeze-drying cycles. *Drying Technol.* 37:579–599.

Pisano, R., Ferri, G., Fissore, D., and A. A. Barresi. 2017. Freeze-drying monitoring via Pressure Rise Test: The role of pressure sensor dynamics. In *Proceedings of IEEE International Instrumentation and Measurements Technology Conference "I2MTC 2017"*, Torino (Italy), May 22–25, 2017, pp. 1282–1287.

Pisano, R., Fissore, D., and A. A. Barresi. 2011a. Freeze-drying cycle optimization using Model Predictive Control techniques. *Ind. Eng. Chem. Res.* 50:7363–7379.

Pisano, R., Fissore, D., and A. A. Barresi. 2011b. Innovation in monitoring food freeze drying. *Drying Technol.* 29:1920–1931.

Pisano, R., Fissore, D., and A. A. Barresi. 2011c. Heat transfer in freeze-drying apparatus. In *Developments in heat transfer*, ed. M. A. dos Santos Bernardes, Chap. 6, 91–114. Rijeka, Croatia: InTech. www.intechopen.com/books/show/title/developments-in-heat-transfer

Pisano, R., Fissore, D., and A. A. Barresi. 2012. Quality by design in the secondary drying step of a freeze-drying process. *Drying Technol.* 30:1307–1316.

Pisano, R., Fissore, D., and A. A. Barresi. 2013a. In-line and off-line optimization of freeze-drying cycles for pharmaceutical products. *Drying Technol.* 31:905–919.

Pisano, R., Fissore, D., and A. A. Barresi. 2014. A new method based on the regression of step response data for monitoring a freeze-drying cycle. *J. Pharm. Sci.* 130:1756–1765.

Pisano, R., Fissore, D., and A. A. Barresi. 2016. Noninvasive monitoring of a freeze-drying process for *tert*-butanol/water cosolvent-based formulations. *Ind. Eng. Chem. Res.* 55:5670–5780.

Pisano, R., Fissore, D., Barresi, A. A., Brayard, P., Chouvenc, P., and B. Woinet. 2013c. Quality by Design: Optimization of a freeze-drying cycle via design space in case of heterogeneous drying behavior and influence of the freezing protocol. *Pharm. Dev. Tech.* 18:280–295.

Pisano, R., Fissore, D., Velardi, S. A., and A. A. Barresi. 2010. In-line optimization and control of an industrial freeze-drying process for pharmaceuticals. *J. Pharm. Sci.* 99:4691–4709.

Pisano, R., Rasetto, V., Barresi, A. A., Kuntz, F., Aoude-Werner, D., and L. Rey. 2013b. Freeze-drying of enzymes in case of water-binding and non-water-binding substrates. *Eur. J. Pharm. Biopharm.* 85:974–983.

Rasetto, V., Marchisio, D. L., Fissore, D., and A. A. Barresi. 2010. On the use of a dual-scale model to improve understanding of a pharmaceutical freeze-drying process. *J. Pharm. Sci.* 99:4337–4350.

Rey, L. R. 1976. Glimpses into the fundamental aspects of freeze-drying. *Dev. Biol. Standard.* 36:19–27.

Rieutord, L. M. A. 1962. Apparatus for regulating freeze-drying operations. U. S. Patent US3192643 filed January 12, 1962.

Rosas, J. G., de Waard, H., De Beer, T. et al. 2014. NIR spectroscopy for the in-line monitoring of a multicomponent formulation during the entire freeze-drying process. *J. Pharm. Biomed. Anal.* 97:39–46.

Sadikoglu, H., 2005. Optimal control of the secondary drying stage of freeze drying of solutions in vials using variational calculus. *Drying Technol.* 23:33–57.

Sadikoglu, H., Liapis, A. I., and O. K. Crosser. 1998. Optimal control of the primary and secondary drying stages of bulk solution freeze drying in trays. *Drying Technol.* 16:399–431.

Sadikoglu, H., Ozdemir, M., and M. Seker. 2003. Optimal control of the primary drying stage of freeze drying of solutions in vials using variational calculus. *Drying Technol.* 21:1307–1331.

Schneid, S., and H. Gieseler. 2008. Evaluation of a new wireless temperature remote interrogation system (TEMPRIS) to measure product temperature during freeze drying. *AAPS PharmSciTech* 9:729–739.

Schneid, S. C., Gieseler, H., Kessler, W. J., and M. J. Pikal. 2009. Non-invasive product temperature determination during primary drying using tunable diode laser absorption spectroscopy. *J. Pharm. Sci.* 98:3406–3418.

Searles, J. A., Aravapalli, S., and C. Hodge. 2017. Effects of chamber pressure and partial pressure of water vapor on secondary drying in lyophilization. *AAPS PharmSciTech* 18:2808–2813.

Sennhenn, B., Gehrmann, D., and A. Firus. 2003. Freeze-drying apparatus. U. S. Patent US6931754 B2 filed April 10, 2003.

Tang, X. C., Nail, S. L., and M. J. Pikal. 2005. Freeze-drying process design by manometric temperature measurement: Design of a smart freeze-dryer. *Pharm. Res.* 22:685–700.

Tchessalov, S. A., and N. W. Warne. 2007. Lyophilization methods and apparatuses. U. S. Patent Application US2008/0098614 A1 filed October 3, 2007.

Tchessalov, S., and N. Warne. 2008. Lyophilization: Cycle robustness and process tolerances, transfer and scale up. *Eur. Pharm. Rev.* 13:76–83 (article 1263).

Tenedini, K. J., and D. T. Sutherland. 1984. Freeze dryer with improved temperature control. U. S. Patent US4547977 A filed May 21, 1984.

Tenedini, K. J., and S. S. Jr. Bart. 1999. Freeze drying methods employing vapor flow monitoring and/or vacuum pressure control. U. S. Patent US6226997 B1 filed December 16, 1999.

Thompson, T. N. 2013. LyoPAT™. Real-time monitoring and control of the freezing and primary drying stages during freeze-drying for improved product quality and reduced cycle times. *Am. Pharm. Rev.* 16:68–74.

Trelea, I. C., Passo, S., Fonseca, F., and M. Marin. 2007. An interactive tool for the optimization of freeze-drying cycles based on quality criteria. *Drying Technol.* 25:741–751.

Velardi, S. A., Rasetto, V., and A. A. Barresi. 2008. Dynamic Parameters Estimation method: Advanced Manometric Temperature Measurement approach for freeze-drying monitoring of pharmaceutical solutions. *Ind. Eng. Chem. Res.* 47:8445–8457.

Velardi, S. A., Hammouri, H., and A. A. Barresi. 2009. In line monitoring of the primary drying phase of the freeze-drying process in vial by means of a Kalman filter based observer. *Chem. Eng. Res. Des.* 87:1409–1419.

Velardi, S. A., Hammouri, H., and A. A. Barresi. 2010. Development of a high gain observer for inline monitoring of sublimation in vial freeze-drying. *Drying Technol.* 28:256–268.

Vollrath, I., Pauli, V., Friess, W., Freitag, A., Hawe, A., and G. Winter. 2017. Evaluation of heat flux measurement as a new process analytical technology monitoring tool in freeze drying. *J. Pharm. Sci.* 106:1249–1257.

Willemer, H. 1987. Additional independent process control by process sampling for sensitive biomedical products. In *Proceedings of 17ème congrès international du froid*, Wien (Austria); Volume C, 146–152.

7 Use of a Micro Freeze-Dryer for Developing a Freeze-Drying Process

Taylor N. Thompson and Davide Fissore

CONTENTS

7.1 INTRODUCTION

The number of new drugs developed by pharmaceutical companies and research laboratories and ready to go on the market is increasing day by day. Quite often it is necessary to remove the water when these drugs are obtained as liquid solutions, with the goal of increasing their shelf-life, and freeze-drying is used to this purpose as these molecules are usually thermolabile (Pikal 1994; Fissore 2013). One of the most challenging case studies is that of biopharmaceuticals (e.g., nucleotide delivery systems, antibody derived binding molecules, etc.), as they represent a large fraction of these new drugs (Otto et al. 2014; Lyophilization Services for Biopharmaceuticals 2017) and freeze-drying is the preferred method to stabilize them.

In this scenario, the need for the reduction of the time to market is the main driving force for developing and implementing new technical solutions that allow accelerating the freeze-drying process design stage.

Several issues must be faced and solved. When using the experimental approach to process design and optimization, many runs are required, both in the case where a design of experiments is used and in the case where a trial-and-error approach is at the basis of the investigation. This is because several tests are required to assess the effect of the operating conditions, namely the temperature of the heating shelf and the pressure in the drying chamber, on drying duration and on maximum product temperature, the main critical process parameters. It is not only necessary to identify the optimal operating conditions, which allow minimizing process duration without jeopardizing product quality, but also to assess their robustness, that is, to identify the design space of the process. Even when using mathematical modeling for optimizing the process, several tests are required to get the values of model parameters and, then, for model validation (Giordano et al. 2011; Fissore et al. 2011a; Koganti et al. 2011; Pisano et al. 2013); model parameters may be obtained also in-line, as discussed in Chapter 6. In all cases, in addition to the duration of the freeze-drying run, the time required to prepare the batch, to load/unload it, and to defrost the condenser has to be considered and represents an important issue that increases the time required to develop the freeze-drying process for a given product. A further issue is the consumption of the active pharmaceutical ingredient, which may be expensive and even not available in sufficient quantity at this stage. Or, the formulation may be not yet be fully developed, and this adds a degree of freedom to the study.

7.2 SMALL-SCALE FREEZE-DRYERS

Small-scale freeze-dryers represent a technical solution that addresses the problems listed above (Gieseler and Gieseler 2017). In fact, in a small-scale freeze-dryer just a few vials are required, which (i) minimizes the amount of the active pharmaceutical ingredient required and (ii) reduces the time needed for batch preparation, loading/ unloading, condenser defrosting, and so on.

The MicroFD® by Millrock Technology, Inc. (Kingston, NY, USA), is one of the technical solutions available on the market to analyze and optimize a freeze-drying cycle using a small number of vials (Thompson et al. 2017). Figure 7.1 show a cylindrical drying chamber, with a 6 in (152 mm) diameter shelf where the vials are loaded: the main feature of the system is LyoSim®, a temperature-controlled ring with removeable thermal-conducting blocks, in direct contact with the external row of vials of the batch, whose size is dependent on the diameter of the vials loaded

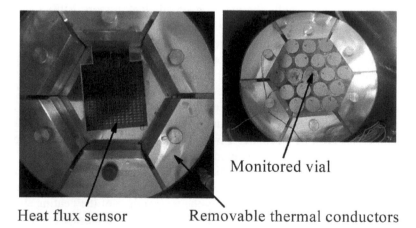

Monitored vial

Heat flux sensor   Removable thermal conductors

FIGURE 7.1   Drying chamber of the MicroFD® with the heat flux sensor (Accuflux®) (LHS) and an example of vials loading in the system (RHS).

in the system (as it will be shown, it is necessary to guarantee an intimate contact between the external vials of the batch and these elements).

When carrying out a freeze-drying cycle in a small-scale unit, the goal is to gather knowledge about the process for use in a large-scale unit. This means, for the selected operating conditions (temperature of the heating shelf and pressure in the drying chamber), to get values of drying duration and product temperature representative of what will occur in a different unit. In this framework, a critical issue is represented by the way the heat required for ice sublimation is transferred to the product. It is in fact well known that heating conditions are not uniform in a batch (Gan et al. 2005; Barresi et al. 2010; Pisano et al. 2011, 2013; Fissore 2013; Pikal et al. 2016), and vials at the edge of the batch, not surrounded by other vials as those in central position, dry faster and the product they process is at a higher temperature with respect to the rest of the batch. This is generally assumed to be because the radiation from chamber walls (and door), which represents an additional heat transfer mechanism to the product. While in a large-scale unit these edge vials represent a small fraction of the batch (usually lower than 5%), in a small-scale unit the situation is quite different: as an example, considering the 19 6R vials loaded in the MicroFD® shown in Figure 7.1, there are 12 "edge" vials and only 7 "central" vials. Therefore, a small batch would normally process the product in a manner that is not representative of larger units.

As an example, Figure 7.2 shows the ratio between the pressure signals obtained through a capacitance manometer (Baratron type) and a thermal conductivity

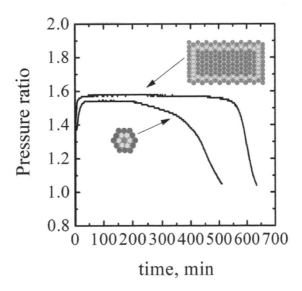

FIGURE 7.2 Comparison between the ratio of signals of the thermal conductivity gauge and that of the capacitance manometer measured in the large-scale freeze-dryer (REVO®, Millrock Technology, Inc.) with a full-tray batch (336 vials) and a small-scale batch (19 vials). Ten-milliliter vials have been used, with a filling volume of 3 mL, processed at 20°C and 133 μbar.

pressure sensor (Pirani type), used to assess the ending point of the primary drying stage (Patel et al. 2010), in two cases: in the first one 336 vials are loaded in a large-scale unit (REVO® by Millrock Technology, Inc.), while in the second only 19 vials are loaded, in the same dryer, in both cases using a hexagonal array.

Drying time estimated through the pressure ratio, at the instant when it falls below a given threshold, is 636 minutes for the full tray test and 512 minutes for the small batch, for the selected operating conditions. This test has been carried out also in the MicroFD®, without using the LyoSim®, and results are shown in Figure 7.3: drying time is 532 minutes, close to the value of 512 minutes obtained in the REVO® freeze-dryer with the same batch, but very far from the 636 minutes of the full-tray batch.

As radiation is usually assumed to be the main cause of nonuniform heat transfer to a batch of vials, to remove this effect or, at least, to minimize its contribution, the test has been repeated in the MicroFD® with the chamber wall cooled to a very low temperature (−20°C). Results are shown in Figure 7.4.

Drying time is now 557 minutes, demonstrating that the control of radiation from chamber walls only results in a small improvement of the batch dynamics, which is now only 25 minutes slower, and does not reach the value required to have equivalence between the small- and the large-scale units. Obeidat et al. (2018) proposed a couple of prototypes where the temperature of the chamber walls, in the first prototype, or that of a cylindrical wall introduced in the chamber, in the second prototype, was manipulated, aiming to get a uniform drying behavior in the batch, close to that of a large-scale unit: also in this case the "edge

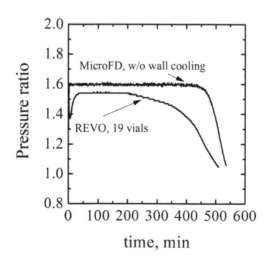

FIGURE 7.3 Comparison between the ratio of signals of the thermal conductivity gauge and that of the capacitance manometer measured in the large-scale freeze-dryer (REVO®, Millrock Technology, Inc.) with a small-scale batch (19 vials) and in a small-scale freeze-dryer (MicroFD®, Millrock Technology, Inc.) without any temperature control of the wall. Ten-milliliter vials have been used, with a filling volume of 3 mL, processed at 20°C and 133 μbar.

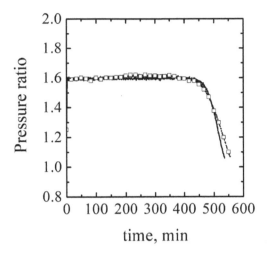

FIGURE 7.4 Comparison between the ratio of signals of the thermal conductivity gauge and that of the capacitance manometer measured in a small-scale freeze-dryer (MicroFD®, Millrock Technology, Inc.) with a small-scale batch (19 vials) without any temperature control of the wall (solid line) and with cooled wall at –20°C (symbols). Ten-milliliter vials have been used, with a filling volume of 3 mL, processed at 20°C and 133 μbar.

vial" effect appears not to have been removed, and the uniformity of product temperature in the batch appeared to be inconsistent (it has to be highlighted that in this system only 7 vials are loaded and 6 of them, i.e., 86% of the batch, are edge vials, with just 1 central vial).

Scutellà et al. (2017, 2018) pointed out that in the external vials of the batch the conduction in the gas surrounding the vials is responsible for the atypical heating conditions, with respect to the central vials, much more than radiation from chamber walls. The LyoSim® used in the MicroFD® aims to imitate an additional row of vials, in contact with the vials of the batch, in such a way that the drying conditions in the external vials are as close as possible to those of the central vials. The temperature of LyoSim® is set based on the product temperature measured in some of the vials of the batch, aiming to maintain a constant difference (offset) between the temperature of the ring and that of the product. From a theoretical point of view, the offset value should be set equal to zero, if the goal is to reproduce in the MicroFD® the dynamics of the central vials of a larger scale unit, which means that the ring temperature should be the same as the product in the vials. Nevertheless, negative values of the temperature offset provide better results, which means that LyoSim® (slightly) cools the external vials of the batch: this is required to compensate for other heating effects that may play some roles. Figure 7.5 shows the results obtained in the MicroFD® with LyoSim® cooling: it appears that the ending time determined from the pressure ratio curves in this system and in the large-scale unit at full load are very close, 633 vs. 636 minutes, thus proving that LyoSim® is able to reproduce in a small batch the behavior of a larger-scale unit.

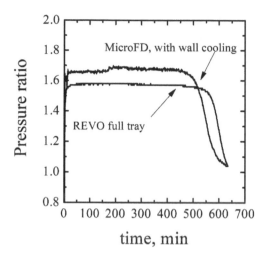

FIGURE 7.5 Comparison between the ratio of signals of the thermal conductivity gauge and that of the capacitance manometer measured in the large-scale freeze-dryer (REVO® Millrock Technology, Inc.) with a full-tray batch (336 vials) and in a small-scale freeze-dryer (MicroFD®, Millrock Technology, Inc.) with LyoSim® temperature control. Ten-milliliter vials have been used, with a filling volume of 3 mL, processed at 20°C and 133 µbar.

7.3   PROCESS INVESTIGATION IN A MICRO FREEZE-DRYER

One of the key issues when using the micro freeze-dryer is the selection of the temperature offset. Figure 7.6 shows the comparison between the mean weight loss in the external and in the internal vials as a function of the temperature offset, after 6 hours of drying of a 5% w/w sucrose solution in the MicroFD®.

In this device, it is in fact possible to manipulate the temperature offset in the range from −15°C to +15°C with respect to the mean product temperature measured through T-type thin thermocouples inserted in some vials of the batch. Negative values of the offset, given as difference between the temperature of the ring and that of the product, are used when we aim to simulate the dynamics of the central vials of the batch. In this case the ring simulates the presence of additional rows of vials where ice sublimation occurs. It is important to note that the LyoSim® temperature offset is set to simulate a temperature somewhere between the product temperature and the temperature of the sublimation front. Positive values of the offset are used when the goal of the investigation is to simulate, in the small-scale unit, the behaviour of the edge vials: as in this case additional heat has to be supplied to the vials, and the temperature of the ring has to be higher than that of the product in the batch. For the case study investigated, whose results are shown in Figure 7.6, it appears that very close values of weight loss, that is, of sublimation flux, are obtained in the batch whichever temperature offset is selected. Considering the batch as a whole, the standard deviation of the weight loss (i.e., of the sublimation flux) ranges from 7.7% for a temperature offset of −1°C, to 5.23% for a temperature offset of −5°C; these values are in the range usually obtained in

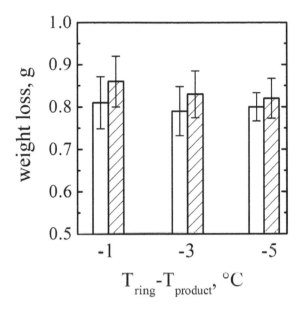

FIGURE 7.6 Comparison between the weight loss in the edge (striped bars) and in the internal (white bars) vials for different values of temperature offset (given as difference between the temperature of the ring and that of the monitored vials) in the small-scale freeze-dryer (MicroFD®, Millrock Technology, Inc.). Ten-milliliter vials have been used, with 3 mL of a 5% w/w sucrose solution, processed at −20°C and 133 μbar; duration of the test was set at 6 hours.

a large-scale batch for central vials. On further review of the data, the difference between the weight losses in the external and in the internal vials is lowest for a temperature offset of −5°C. This type of study is very fast, as it is not required to carry out the primary drying stage until the end, and, thus, few days are needed to optimize the value of the temperature offset.

In addition to the rate of sublimation, product temperature is the other critical parameter that must be considered when evaluating the effect of the temperature offset on the batch. Figure 7.7 compares the temperature over a period of time in one of the central vials of the batch with that of one of the edge vials in the case of two values of temperature offset. In this case it appears that the best results are obtained with a temperature offset of −5°C, where the two curves are almost overlapping.

7.4 MODEL PARAMETERS ESTIMATION FOR PROCESS OPTIMIZATION

When the goal of the study is the optimization of the process, it is necessary to estimate in the small-scale freeze-dryer the value of the model parameters for *in silico* process simulation and optimization for the large-scale unit. Several mathematical models were proposed in the past to simulate the primary drying stage, the most

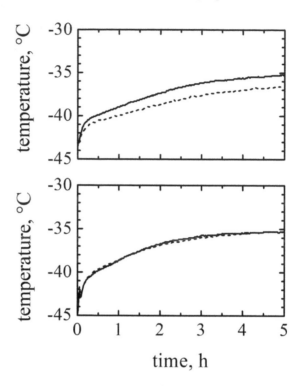

FIGURE 7.7 Comparison between the temperature evolution in the edge (solid lines) and in the central (dashed lines) vials for different values of temperature offset (graph A, $-3°C$; graph B, $-5°C$) when processing a 5% w/w sucrose solution in a small-scale freeze-dryer (MicroFD®, Millrock Technology, Inc.). Ten-milliliter vials have been used, with a filling volume of 3 mL, processed at $-20°C$ and 80 μbar.

critical part of a freeze-drying process (see, among the others, Fissore et al. 2015). One-dimensional models, neglecting the radial gradients of temperature and composition and, thus, assuming a flat interface of sublimation receding from the top of the vial to the bottom, were shown to provide accurate values of product temperature and drying duration (Velardi and Barresi 2008) involving just two parameters, namely the heat transfer coefficient K_v and the resistance of the dried product to vapor flux R_p. K_v is used to express the dependence of the heat flux to the product, J_q, on the driving force given by the difference between the temperature of the heating shelf, T_s, and that of the product at the bottom of the vial, T_B:

$$J_q = K_v \left(T_s - T_B \right).$$ (7.1)

R_p is used to express the dependence of the sublimation flux, J_w, on the driving force given by the difference between water vapor partial pressure at the interface of sublimation, $p_{w,i}$ (that is a well-known function of product temperature, see, among the others, Fissore et al. 2011b), and in the drying chamber, $p_{w,c}$ (that can be assumed to be equal to total chamber pressure, because it is the atmosphere in the chamber composed almost completely by water vapor):

$$J_w = \frac{1}{R_p}\left(p_{w,i} - p_{w,c}\right).$$

(7.2)

The model includes the following equations:

— the energy balance at the interface of sublimation:

$$J_q = \Delta H_s J_w,$$

(7.3)

where ΔH_s is the heat of sublimation;

— the mass balance for the frozen layer:

$$\frac{dL_f}{dt} = -\frac{1}{\rho_f - \rho_d}J_w,$$

(7.4)

where L_f is the thickness of the frozen product, t is the time, ρ_f and ρ_d are, respectively, the density of the frozen and of the dried product;

— the energy balance for the frozen product:

$$T_B = T_s - \frac{1}{K_v}\left(\frac{1}{K_v} + \frac{L_f}{k_f}\right)^{-1}(T_s - T_i),$$

(7.5)

where k_f is the thermal conductivity of the frozen product and T_i the temperature of the interface of sublimation.

With respect to the heat transfer coefficient K_v, several process analytical technologies have been proposed even in recent years to allow a fast estimation of this parameter (see the recent review by Fissore et al. 2018), whose value is dependent, mainly, on the type of vials and on the pressure in the drying chamber. K_v will also vary between freeze-dryers owing to the variation in heat sources, shelf surface finishes, mass of the shelf, heat transfer fluid, and fluid flow rates inside the shelf.

The MicroFD® is equipped with a heat flux sensor, AccuFlux®, shown in Figure 7.1: it is a thin-film differential thermopile placed in contact with the bottom of the vials and with the shelf, producing an electrical signal proportional to the heat flux through the surface, thus allowing a direct measurement of the heat flux from the shelf to the product (Ling 2013; Vollrath et al. 2017). AccuFlux® allows direct measurement of K_v using Equation 7.1 where T_s and T_B are measured. For the case study presented in this chapter, that is, the drying of a 5% w/w sucrose solution in 10 mL vials, at 80 µbar and −20°C, values of 6.79, 6.73, and 6.53 W m^{-2}K^{-1} are obtained for a temperature offset of $-1°C$, $-3°C$, and $-5°C$, respectively. Very close values are obtained in the three cases as, evidently, the ring temperature does not affect the heat flux provided by the heating shelf.

It has to be highlighted that AccuFlux® measures the heat transferred through the shelf, which is not the parameter to be used in Equation 7.1, and it should thus be indicated as $K_{v,shelf}$, with the heat flux measured indicated as $J_{q,shelf}$. The coefficient K_v has in fact to be regarded as an effective heat transfer coefficient that accounts for all the heat transfer mechanisms to the product. In order to determine its value, it is possible to use the weight loss measured in the preliminary test carried out to assess the effect of the temperature offset. In fact, being Δm the weight loss measured in one of the vials, whose cross area is A_v, in the test whose duration is t_d, the coefficient K_v can be calculated using the following equation:

$$K_v = \frac{\Delta m \Delta H_s}{A_v \int_0^{t_d} (T_s - T_B) dt}, \tag{7.6}$$

Equation 7.6 is obtained from the energy balance for the frozen product, assuming that all the heat transferred to the product is used for ice sublimation. In our case study, values of 15.8, 15.61, and 15.26 W m^{-2}K^{-1} are obtained for temperature offset of the ring equal to $-1°C$, $-3°C$, and $-5°C$, respectively.

As an alternative, if a full drying cycle is carried out, at the end of the test K_v can be calculated using again Equation 7.6, considering that Δm is the amount of water poured in each vial at the beginning of the test.

With respect to the heat transfer coefficient K_v, by modifying the temperature of the ring it is possible to get very different values of K_v, as it was recently investigated by Goldman et al. (2019); by this way it becomes possible to replicate, in the small-scale unit, the dynamics not only of central vials of a different unit but also that of the edge vials, by properly tuning the value of K_v. When more than 7 vials are being processed, the shelf temperature may be adjusted to adjust the batch K_v to mimic larger systems.

The coefficient R_p, for a given product, is a function of the freezing protocol and of the thickness of the dried product, L_d. Thus, to achieve in the micro freeze-dryer values of R_p vs. L_d representative of a large-scale unit, the freezing protocol must be the same [in the MicroFD® it is also possible to carry out controlled nucleation using the ice-fog technique (Ling 2011)]. Then, it is required to know K_v, and thus, it is possible to get the curve of R_p vs. L_d at the end of the primary drying, when K_v is known, or during a run if K_v has been previously determined. In this latter case the procedure is very simple and is summarized in the following:

1. At $t = 0$, $L_d = 0$ and, thus, $R_p = 0$.
2. After a period of time Δt it is possible to calculate both the amount of heat transferred to the product

$$Q = K_v A_v \int_t^{t+\Delta t} (T_s - T_B) dt \tag{7.7}$$

and the mean heat flux

$$J_q = \frac{Q}{A_v \Delta t}.$$ (7.8)

3. The mass flux can then be calculated using Equation 7.3.
4. R_p is finally calculated through Equation 7.2 with the measured product temperature and assuming that the axial gradient of temperature in the frozen product is negligible.
5. The variation of L_d can be calculated by assuming that the total thickness of the product does not change and that the thickness of the frozen layer decreases according to Equation 7.4.
6. All previous calculations have to be repeated for the following time interval to obtain the curve of R_p vs. L_d.

Figure 7.8 shows the curve of R_p vs. L_d obtained for the 5% w/w sucrose solution processed at −20°C and 80 μbar and considering two different temperature measurements; two almost overlapping series of values are obtained, as a consequence of the fact that the ice sublimation in the two monitored vials exhibits a very similar thermal history.

Using the calculated values, it is possible to calculate the parameters ($R_{p,0}$, A, and B) of the function generally used to express the dependence of R_p on L_d:

$$R_p = R_{p,0} + \frac{A \cdot L_d}{1 + B \cdot L_d}.$$ (7.9)

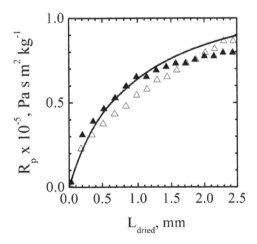

FIGURE 7.8 Comparison between the values of R_p vs. L_{dried} calculated using two temperature measurements (in two different vials); solid line identifies the curve calculated using Equation 7.9 and looking for the best fit between calculated and measured values. (Operating conditions: $T_s = -20°C$, $P_c = 80$ μbar, 5% w/w sucrose solution, 10 mL vials, 3 mL filling volume).

Through least-squares minimization, we get $R_{p,0} = 1 \times 10^4$ m s^{-1}, A = 9.1 $\times$ 10^7 s^{-1} and B = 7.2 $\times$ 10^2 m^{-1}.

Once model parameters have been estimated it becomes possible to use mathematical modeling for evaluating the design space of the process (Giordano et al. 2011; Fissore et al. 2011a; Koganti et al. 2011) and, thus, for optimizing the process.

7.5 CONCLUSION

Results presented in this chapter demonstrate that the LyoSim® concept implemented in the MicroFD® is effective to achieve uniform drying behavior in a small-scale batch. The temperature of the metallic ring in contact with the external vials of the batch may be optimized by carrying out few tests to reduce the nonuniformity of the drying rate and product temperature among the vials. A few gravimetric tests are enough to get the desired result. In addition, through the AccuFlux® sensor, it is possible to evaluate the heat flux from the shelf to the product and, then, using the sequence of calculations described in detail in section 7.4, it is possible to calculate the parameters of a simple one-dimensional mathematical model that can then be used for off-line process simulation and optimization.

ACKNOWLEDGMENTS

The authors would like to acknowledge Giuseppe Gallo and Angelo Emiliano Ruggiero (Politecnico di Torino), for their contribution to the experimental investigation, and Stefano Montagnoli and Paolo Sanesi (Pharmatech, Italy) for their technical support and valuable suggestions.

REFERENCES

Barresi, A. A., Pisano, R., Rasetto, V., Fissore, D., and D. L. Marchisio. 2010. Model-based monitoring and control of industrial freeze-drying processes: Effect of batch nonuniformity. *Drying Technol.* 28:577–590.

Fissore, D. 2013. Freeze-drying of pharmaceuticals. In *Encyclopedia of pharmaceutical science and technology,* 4th Edition, ed. J. Swarbrick, 1723–1737. London: CRC Press.

Fissore, D., Pisano, R., and A. A. Barresi. 2011a. Advanced approach to build the design space for the primary drying of a pharmaceutical freeze-drying process. *J. Pharm. Sci.* 100:4922–4933.

Fissore, D., Pisano, R., and A. A. Barresi. 2011b. On the methods based on the Pressure Rise Test for monitoring a freeze-drying process. *Drying Technol.* 29:73–90.

Fissore, D., Pisano, R., and A. A. Barresi. 2015. Using mathematical modeling and prior knowledge for QbD in freeze-drying processes. In *Quality by design for biopharmaceutical drug product development,* ed. F. Jameel, S. Hershenson, M. A. Khan, and S. Martin-Moe, 565–593. New York: Springer.

Fissore, D., Pisano, R., and A. A. Barresi. 2018. Process analytical technology for monitoring pharmaceuticals freeze-drying – A comprehensive review. *Drying Technol.* 36:1839–1865.

Gan, K. H., Bruttini, R., Crosser, O. K., and A. A. Liapis. 2005. Freeze-drying of pharmaceuticals in vials on trays: Effects of drying chamber wall temperature and tray side on lyophilization performance. *Int. J. Heat Mass Tran.* 48:1675–1687.

Gieseler, H., and M. Gieseler. 2017. The importance of being small: Miniaturization of freeze-drying equipment. *Eur. Pharm. Rev.* 4:28–32.

Giordano, A., Barresi, A. A., and D. Fissore. 2011. On the use of mathematical models to build the design space for the primary drying phase of a pharmaceutical lyophilization process. *J. Pharm. Sci.* 100:311–324.

Goldman, J. M., Chen, X., Register, J. T., et al. 2019. Representative scale-down lyophilization cycle development using a seven-vial freeze-dryer (MicroFD®). *J. Pharm. Sci.* 108:1486–1495.

Koganti, V. R., Shalaev, E. Y., Berry, M. R., et al. 2011. Investigation of design space for freeze-drying: Use of modeling for primary drying segment of a freeze-drying cycle, *AAPS PharmSciTech* 12:854–861.

Ling, W. 2011. Controlled nucleation during freezing step of freeze-drying cycle using pressure differential ice fog distribution. U. S. Patent Application 2012/0272544 A1 filed April 29, 2011.

Ling, W. 2013. Using surface heat flux measurement to monitor and control a freeze-drying process. U. S. Patent 9121637 B2 filed June 25, 2013.

Lyophilization Services for Biopharmaceuticals, 2017–2027. 2017. www.reportbuyer.com/product/4886788/lyophilization-services-for-biopharmaceuticals-2017-2027.html (accessed January 2019).

Obeidat, W. M., Bogner, R., Mudhivarthi, V., Sharma, P., and P. Sane. 2018. Development of a mini-freeze dryer for material-sparing laboratory processing with representative temperature history. *AAPS PharmSciTech* 29:599–609.

Otto, R., Santagostino, A., and U. Schrader. 2014. Rapid growth in biopharma: Challenges and opportunities. www.mckinsey.com/industries/pharmaceuticals-and-medical-products/our-insights/rapid-growth-in-biopharma (accessed January 2019).

Patel, S. M., Doen, T., and M. J. Pikal. 2010. Determination of end point of primary drying in freeze-drying process control. *AAPS PharmSciTech* 11:73–84.

Pikal, M. J. 1994. Freeze-drying of proteins: Process, formulation, and stability. In *Formulation and delivery of proteins and peptides*, ed. J. L. Cleland, and R. Langer, 120–133. Washington, DC: American Chemical Society.

Pikal, M. J., Bogner, R., Mudhivarthi, V., Sharma, P., and P. Sane. 2016. Freeze-drying process development and scale-up: Scale-up of edge vial versus center vial heat transfer coefficients, K_v. *J. Pharm. Sci.* 105:3333–3343.

Pisano, R., Fissore, D., and A. A. Barresi. 2011. Heat transfer in freeze-drying apparatus. In *Developments in heat transfer*, ed. M. A. Dos Santos Bernardes, 91–114. Rijeka: Intech.

Pisano, R., Fissore, D., Barresi, A. A., Brayard, P., Chouvenc, P., and B. Woinet. 2013. Quality by Design: Optimization of a freeze-drying cycle via design space in case of heterogeneous drying behavior and influence of the freezing protocol. *Pharm. Dev. Tech.* 18:280–295.

Scutellà, B., Bourlès, E., and A. Plana-Fattori. 2018. Effect of freeze dryer design on heat transfer variability investigated using a 3D mathematical model. *J. Pharm. Sci.* 107:2098–2106.

Scutellà, B., Plana-Fattori, A., Passot, S., et al. 2017. 3D mathematical modeling to understand atypical heat transfer observed in vial freeze-drying. *Appl. Therm. Eng.* 12:226–236.

Thompson, T. N., Wang, Q., and C. Reiter. 2017. Developing transferable freeze-drying protocols using Accuflux® and a MicroFD®. Presentation given at Peptalk 2017, San Diego, USA. https://pharmahub.org/groups/lyo/tools (accessed January 2019).

Velardi, S. A., and A. A. Barresi. 2008. Development of simplified models for the freeze-drying process and investigation of the optimal operating conditions. *Chem. Eng. Res. Des.* 86:9–22.

Vollrath, I., Pauli, V., Friess, W., Freitag, A., Hawe, A., and G. Winter. 2017. Evaluation of heat flux measurement as a new process analytical technology monitoring tool in freeze drying. *J. Pharm. Sci.* 106:1249–1257.

8 Continuous Manufacturing in Lyophilization of Pharmaceuticals
Drawbacks of Batch Processing, Current Status, and Perspectives

Roberto Pisano, Luigi C. Capozzi, and Jos A.W.M. Corver

CONTENTS

8.1   INTRODUCTION

Pharmaceutical manufacturing often combines batch and continuous unit operations. However, if certain unit operations can be considered continuous, the manufacturing process as a whole is batch in nature. As such, pharmaceutical processes are relatively inefficient and less understood with respect to those in other chemical process industries (Myerson et al. 2015).

In recent years, many pharmaceutical companies have become interested in the development of technologies that have a great deal of potential to improve agility, flexibility, and robustness in the manufacture of drug products. In response to this request, federal agencies have encouraged and sustained advancements in the pharmaceutical manufacturing. One such innovation is continuous manufacturing, where individual continuous unit operations are connected one another to form an integrated manufacturing process. Of course, this manufacturing architecture involves high-level process design and adequate control, since any deviations in the quality of intermediate products may create disruption for downstream unit operations and have an impact on the quality of the products. Advanced process control systems are thus necessary to mitigate any impact of process and raw material variability on the final quality of drug products, and process analytical technologies (PATs) have to be utilized to provide real-time data for process monitoring (Baxendale et al. 2015).

Continuous manufacturing provides several potential opportunities to enhance process efficiency and improve the quality of the finished products. For example, continuous processes generally offer higher throughput per unit time and per unit volume than batch processes and require smaller equipment. For this reason, the application of continuous manufacturing generally results in substantial reduction of both capital and operating expenses. An economic barrier to the adoption of continuous technologies can be given by the current inventory of available batch manufacturing facilities. For new manufacturing strategies to be adopted, there should be a need for significant expansion of manufacturing capacity, or there should be a need for specific process requirements that apply to new biotech products (Poechlauer et al. 2012). Currently, an increasing number of biological drug products are in development or close to entering the market; these drugs are good candidates for the application of new manufacturing strategies. These substances are often highly sensitive to external factors, such as oxygen, heat, sunlight, and pH shift, and most of them have limited stability in liquid form and, hence, are often lyophilized to preserve their potency and prolong their shelf life. There is a large number of pharmaceutical products that are currently lyophilized, including antibiotics, vaccines, bacteria, sera, diagnostic medications, proteins, biotechnological products, high-potency drugs, cells, tissues, and chemicals.

Nonetheless, lyophilization is still considered by many to be both time- and cost-intensive, although it is a well-established and mature technology. Over the past few decades, much effort has been put into understanding the physics of lyophilization and using such an understanding to implement quality-control strategies and advances in equipment design to achieve specific objectives. These innovations, along with the adoption of the quality-by-design (QbD) paradigm and the use of PAT for designing and controlling manufacturing, have resulted in significant progress in terms of cycle duration and control of the critical quality attributes (CQAs) of the product. Notwithstanding these improvements in process efficiency, lyophilization is still characterized by long cycles that are susceptible to failing, and concerns still remain on the vial-to-vial and batch-to-batch heterogeneity. As a matter of fact, the process is still considered to be inefficient, and many companies, regulatory agencies, and universities are striving to improve its efficiency. It is opinion of the authors that some drawbacks of the process can be mitigated by the most recent technologies, such as controlled nucleation, but remain largely unresolved as long as the process is carried out in batch mode. Furthermore, as a result of the rapid development of biopharma industry and the introduction of new drugs and biosimilars, the lyophilization equipment market is forecasted to double its value from \$2.7 billion to \$4.8 billion in 2020, with a compound annual growth rate of 8.5% (Keenan 2014). This scenario opens new opportunities for the development and introduction of more efficient technologies. In general, the transition to continuous lyophilization can be beneficial to process profitability, safety, and improved quality of finished products. Although it is true that there have been advances in science and engineering to support the implementation of continuous manufacturing of lyophilized products, this transition is not straightforward and strongly depends on the collaborative work between academia, regulatory bodies, equipment manufacturers, and of course the pharmaceutical companies. This book chapter summarizes the potential advantages of continuous lyophilization of drug products and gives emphasis to some peculiar quality aspects for consideration and how they may be addressed. It has been tailored for biotechnologists, life scientists, drug developers, engineers, and researchers, to give them an update on the current developments on continuous lyophilization and the impact that this technology might have on their research fields.

8.2   THE IMPORTANCE OF BEING CONTINUOUS

The transition from batch to continuous operations is the new challenge of pharmaceutical manufacturing and, hence, of lyophilization as a downstream operation. The driving force for this transition resides on the advantages that continuous manufacturing can give in terms of process efficiency, agility and flexibility, uniformity and better control of the finished product quality, reduction in processing costs and equipment footprint. This transition is the natural evolution of a process started with the application of the QbD paradigm to pharmaceutical development. According to this principle, product quality is not anymore tested in products but is built in by design, and the design of the manufacturing process has to consistently deliver the intended performance of the products. The first result of this new manufacturing mindset has been the development of new PATs to support real-time monitoring and automatic control of the process, and pharmaceutical quality systems (PQS) and real-time release testing (RTRT) to ensure the quality of finished products. The pharmaceutical community has recently recognized that continuous manufacturing will better fulfill the requirements set by the regulatory authorities (Allison et al. 2015; Nasr et al. 2017).

A similar trend has been observed for lyophilization. Over the past two decades, the lyophilization community has worked on the development of advanced process control systems and improved development and scale-up procedures that should deliver the intended quality and uniformity of the finished products (Lim et al. 2016; Goshima et al. 2016). Of course, this activity was supported by the development and application of new PATs (Fissore et al. 2019) such as software sensors (Velardi et al. 2010; Bosca et al. 2017), wireless temperature sensors (Oddone et al. 2015), infrared thermography (Van Bockstal et al. 2018a; Lietta et al. 2019), LyoFlux and TDLAS (Lackner 2007), Fiber Bragg grating (Al-Fakih et al. 2012), and miniature spectrometers (De Beer et al. 2009; Vollrath et al. 2017). Despite these continuous improvements, lyophilization as a batch-wise process still presents serious limitations, as discussed in section 8.4. Fundamentally, most of these efforts relate to information on the whole batch without reference to individual vials.

8.3   REGULATORY PERSPECTIVES

The implementation of continuous manufacturing into the pharmaceutical industry has been widely promoted and encouraged by the U.S. Food and Drug Administration (FDA), the European Medicines Agency (EMA) and the Japanese Pharmaceuticals and Medical Devices Agency (PMDA) as a way to enhance the level of product quality and increase the flexibility and agility of the pharmaceutical products. Although there is no specific guidance, the FDA has repeatedly declared that continuous manufacturing is compatible with the current regulatory environment, and in particular within the International Conference for Harmonization (ICH) guidance (ICH Q8, Q9, Q10, and Q11, Points to Consider documents). In February 2019, the FDA released the draft guidance for industry *Quality Considerations for Continuous Manufacturing* with a specific focus on scientific and regulatory considerations related to continuous manufacturing. On many occasions, the EMA declared that continuous manufacturing is compatible with the EU Guidelines on process validation, RTRT, manufacture of drug products, chemistry of new active substances, use of NIR, the GMP Annex 15 and Annex 17, and European Pharmacopoeia (Mihokovic 2017). Moreover, in the most recent documents and guidelines (e.g., *Guideline on Manufacture of the Finished Dosage Form*), continuous manufacturing is explicitly mentioned. The regulatory authorities have also clarified the meaning of *batch* and *lot* in the legislation. The terms *batch* and *lot*, as defined in the US regulations (US Code of Federal Regulations 21 CFR 210.3), are not related to a specific mode of manufacture but to the amount of material produced in a unit of time or produced so to have uniform character and quality within specified limits. In the EU GMP Guide Part II, batch/lot is specifically defined in the case of continuous manufacturing as a fraction of the production defined either on a fixed-time or fixed-quantity basis.

Although the current regulatory framework is fully supportive of continuous manufacturing, several issues might arise in the standard approach used by reviewers/assessors and investigators/inspectors to assess these new manufacturing strategies in the current regulatory framework. In this sense, the FDA, EMA, and PMDA have created multidisciplinary teams specialized in the implementation of emerging technologies (Nasr et al. 2017). Similar to other processes in the pharmaceutical industry, there is no reason for not implementing lyophilization in a continuous mode, if the process is able to respond to the quality and safety requests set by the regulatory authorities.

8.4 DRAWBACKS OF BATCH LYOPHILIZATION

8.4.1 PROCESSING AND DEAD TIMES

Depending on the maximum allowable product temperature, the duration of a lyophilization cycle typically ranges from few days to few weeks. In addition to the processing time, other ancillary operations contribute to the total cycle time:

1. loading and unloading of vials into the drying chamber
2. cleaning-in-place operation (CIP)
3. sterilization-in-place operation (SIP)
4. filter integrity test
5. leak test of the unit
6. venting/backfilling
7. condenser defrosting

All these operations have to meet the GMP guidelines and the stringent requirements set by regulatory authorities for sterile products. The filling of vials, loading, and unloading are performed using automatic systems with a capacity of 10,000–30,000 vials per hour (Oetjen 1999), which means that, for a production of 100,000 vials, loading and unloading can range between 6 and 10 hours. Other auxiliary operations include line clearance, defrosting, CIP, SIP, water intrusion testing, and leak testing, which take between 7 and 13 hours. Thus, the overall downtime has a severe impact on the overall processing time, decreasing the efficiency and profitability of the process.

8.4.2 POOR CONTROL OF THE FREEZING PROCESS

Freezing is a crucial phase of the lyophilization process as it determines the morphology, polymorphism, potency, and stability of finished products (Peters et al. 2016). The critical process parameters (CPPs) of freezing are supercooling or nucleation temperature and freezing rate. The stochastic nature of nucleation makes supercooling to be stochastically distributed and is responsible for both vial-to-vial and batch-to-batch heterogeneity in terms of drying behavior and final product characteristics (Capozzi and Pisano 2018; Oddone et al. 2016; Peters et al. 2016; Searles et al. 2001). These effects can currently be mitigated by the application of the most modern control-freezing technologies, including vacuum induced surface freezing (VISF), ice-fog technique, high-pressure shift freezing, and the depressurization technique (Gasteyer et al. 2015; Oddone et al. 2014; Pisano 2019; Rambhatla et al.

2004). However, most of these technologies have been developed and validated on laboratory-scale equipment, and some concerns still remain on their implementation in the industrial-scale freeze-dryers. Additionally, after nucleation, the freezing is an exothermic process, which leads to vial-to-vial variations due to ill-defined thermal contact with the shelf.

8.4.3 Primary and Secondary Drying

A critical aspect of primary drying is the non-uniformity of heat transfer within a batch of vials, which has implications for both product quality and process performance. During drying, heat is supplied to the product by direct contact between the shelf and the vial, conduction through the gas between the shelf and the bottom of the vial, and radiation. As extensively shown in the literature, a small variation in the vial geometry (Pikal et al. 1984; Scutellà et al. 2017) or the imperfect planarity of the shelf has a great impact on heat transfer. The position of the vial on the shelf represents a further cause of nonuniformity for heat transfer, that is, edge-vial effect. In fact, vials placed at the side of the batch receive much more heat than those placed in the central part (Pikal et al. 2016). Other issues in the batch apparatus are the nonuniformity in the shelf surface temperature and the nonuniformity of pressure within the drying chamber (Alexeenko et al. 2009; Barresi et al. 2010; Rasetto et al. 2010).

The inaccurate control of heat transferred to the product and, thus, of its temperature, seriously impacts the quality of the finished products, as well as the robustness and efficiency of the process. When the product temperature exceeds its maximum allowable value, the product may lose its structure, and the entire batch may be rejected because of unachieved residual moisture levels, the lack of pharmaceutical elegance, or even worse, loss of the API potency (Patel et al. 2017; Wang et al. 2015). Moreover, nonuniformity in the cake morphology and in heat transfer is often responsible of large variability in the residual moisture of the finished products (Pikal and Shah 1997). This variability still remains even if the product structure is better controlled, such as when using controlled nucleation (Gieseler and Stärtzel 2012; Oddone et al. 2017). The control of all these properties is in certain cases fundamental for the long storage of the finished products without losing the API's potency (Breen et al. 2001; Chang et al. 2005).

8.4.4 Batch-to-Batch Variability

Similar to other batch-wise processes, lyophilization has to address batch-to-batch variations in product quality and characteristics. To mitigate these unpredictable variations, one might think that an accurate control of process parameters, chamber pressure, or shelf temperature is enough to maintain consistent quality of the finished products. Unfortunately, changes from batch-to-batch are determined by a multitude of factors, such as variations in grade of ingredients or materials provided by different suppliers, changes in equipment performance and efficiency due to wear, and of course, changes introduced by operators (Galan 2016).

8.4.5   CYCLE SCALE-UP ONTO GMP EQUIPMENT

During preclinical studies, lyophilization is usually carried out in small laboratory equipment, using from a few to a hundred vials per batch. At the end of this phase, the clinical and market production requires bigger apparatus, pilot or commercial scale equipment, to ensure GMP conditions and preserve safety and efficacy of the final product. Scale-up of the cycle plays, at that point, a crucial role in developing a robust and efficient cycle, which can support process validation. The application of QbD approaches to the scale-up of cycles has mitigated its risk of failure, but these approaches often require extensive experimental campaigns and knowledge of the equipment characteristics. The use of mathematical modeling has dramatically reduced the experimental effort (Bosca et al. 2015; Pisano et al. 2013a, 2013b; Van Bockstal et al. 2017b).

8.5   CONTINUOUS LYOPHILIZATION OF PHARMACEUTICALS

The idea of continuous lyophilization dates back to at least the 1940s, when a pilot plant for continuous lyophilization of juice was proposed (Sluder et al. 1947), but only in the 1970s was a fully continuous line, the Conrad freeze-dryer, successfully developed for the production of instant coffee (Goldblith et al. 1975). After this application, the food industry assisted the development of various continuous strategies for the lyophilization of food at industrial, large-scale productions.

Until now, many ideas regarding continuous lyophilization of pharmaceuticals have been proposed (Bruttini 1992; Oetjen et al. 1969; Rey 2010), but none of them has been successfully applied in the industry because of their complexity, the difficulty of guaranteeing product sterility, or handling safety problems. In the continuous apparatus proposed by Rey (2010), the solution is frozen into spherical granules of a given size and loaded continuously onto a heated conveyor in a drying chamber. A vibrating heated tray supplies heat and moves the granules throughout the apparatus; finally, the dried particles are discharged and distributed into vials. This technology can potentially reduce the drying time because the lyophilization of individual granules is very fast compared with frozen liquid solutions or packed beds of particles in vials (Capozzi et al. 2019a, 2019b). However, difficulties can arise from handling small granules and ensuring sterility of the product (Pisano et al. 2015).

In recent years, three continuous technologies are emerging, which aim to address the drawbacks associated with batch lyophilization: lyophilization of granular material proposed by Hosokawa Micron BV (van der Wel 2015), and two distinct configurations for the continuous lyophilization of drug products in unit doses: a first configuration designed by RheaVita in collaboration with Ghent University (Corver 2013), and a second technology proposed by Politecnico di Torino and Massachussetts Institute of Technology (Trout et al. 2018). The main features, pros, and cons of these technologies are discussed below. A more detailed discussion of the various continuous lyophilization technologies that have been patented over the years is given in the article by Pisano et al. (2019).

8.5.1 The Continuous Lyophilization of Drug Products in Bulk

In 2005, the Dutch company Hosokawa Micron BV designed and launched in the market the active freeze-drying (AFD) technology, a system for the production of lyophilized drugs in the form of fine-particle powder (van der Wel 2004). Contrary to classical spray-freeze drying, this process can be performed in batch or continuous/semicontinuous mode, although, to the best of our knowledge, a pilot plant has been tested only in the batch mode (Touzet et al. 2018). As shown in Figure 8.1, the solution is fed into a conical chamber and continuously stirred. The temperature of the solution being frozen is controlled through a heating/cooling jacket. During this stage, the solution solidifies into frozen, macroscopic, blocks of nano/micro crystals. Once freezing has been completed, the pressure is lowered, and heat is supplied from the heating jacket, so that ice crystals in the external layers of the frozen blocks sublimate.

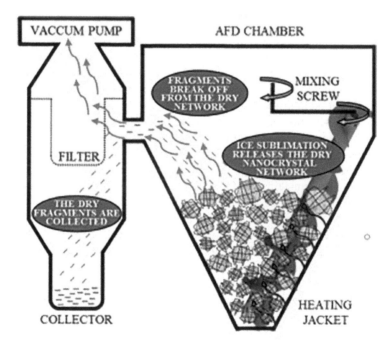

FROZEN BLOCKS OF NANOSUSPENSION

FRAGMENTS OF THE NANOCRYSTAL NETWORK

VAPOR TRANSFER

HEAT TRANSFER

FIGURE 8.1 Schematic representation of AFD technology. Reprinted with permission from Touzet et al. (2018).

Source: Copyright 2018 Elsevier.

At the same time, mixing stresses break the dried layer into small fragments, which are then transported by the vapor flow toward the vacuum system. A filter cartridge is inserted between the tank and the vacuum system, allowing the collection of the dried powder and, finally, its extraction from the equipment. AFD is equipped with SIP/CIP systems and meets aseptic requirements.

This technology is particularly suitable for nanocrystal applications, and, according to the manufacturer, can handle drying solutions, dispersions, pastes, and wet solids. In a recent paper, AFD technology was tested in batch mode for production of nanocrystal-based powder using ketoconazole as a poorly water-soluble model drug (Touzet et al. 2018). Furthermore, van der Wel (2017) showed that AFD can dramatically reduce drying time thanks to the large surface area of the particles and fragments formed during the agitation of the frozen solution, which result in high sublimation area and hence sublimation rate. For example, this technology could reduce the drying time of skimmed milk powder from 50 to 15 h. The AFD manufacturer also claims that the batch-wise process has been successfully tested for a wide range of materials, including nutraceuticals, vitamins, herbal extracts, vegetables, meat products, herbs, milk derivatives, insects, fibers, soups, flavor and broth extracts (van der Wel 2015), but to the best of our knowledge, none of these data have been disclosed. In our opinion, this lack of knowledge and testing in continuous mode might represent severe weaknesses for the diffusion of this process at the industrial level. A similar concept has been patented by IMA Life North America in 2010. This technology uses spray-freezing followed by lyophilization in an agitated tank, which can potentially work in continuous or semicontinuous mode (Demarco and Renzi 2010). Even if this process is adequately executed and the efficacy of the product is as required, there is the challenge of filling the powder in adequately accurate doses. The fact that there is much product handling interaction, leading to challenges in reducing the number of undesired particles, poses another challenge to achieve GMP requirements.

8.5.2 Continuous Lyophilization in Unit Doses, Based on Spin Freezing and IR-Assisted Vacuum Drying

Corver (2013) proposed and filed a patent combining spin-freezing and IR-assisted drying to realize continuous lyophilization of unit doses of drug products. This process concept was then further investigated by the research group of Prof. De Beer at Ghent University (De Meyer et al. 2015; Van Bockstal et al. 2017a, 2018a, 2018b). The technology starts with the spin-freezing of a liquid formulation in a vial, which occurs by rotating the vial along its longitudinal axis and removing heat by flowing a sterile cryogenic gas at controlled temperature and flow rate around it. The partially frozen vial is then transferred into a second, conditioned chamber where the solidification of the solution is completed and annealing may be effectuated. As shown in Figure 8.2, during freezing, the vial is longitudinally rotated at about 2500 rpm; this operation spreads the liquid solution on the vial side wall as uniformly as possible. After that, the vial is transferred into a drying chamber operating at low pressure, where primary and secondary drying occurs. Here, each vial is not in contact with the shelf but slowly rotates at about 5–12 rpm in front of its individual IR heat source,

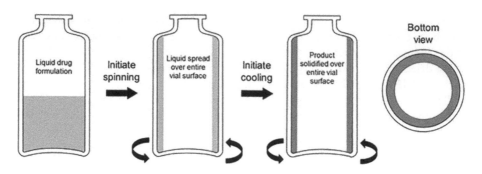

FIGURE 8.2 Spin-freezing of a liquid solution in vials. Reprinted with permission from RheaVita.

which supplies the heat required for ice sublimation. Rotation of the vial is fundamental to achieve a uniform heat transfer from the IR source to the vial.

Accurate control of the product temperature for each vial can then be achieved by adjusting the temperature of the corresponding radiator (Van Bockstal et al. 2017c), while the final residual moisture level can be controlled by adaptively adjusting the power of the IR sources and/or by adjusting the residence time in the drying module. The transfer of vials from chambers at different pressure is achieved by using load-lock systems. Figures 8.3 and 8.4 show a schematic of the continuous lyophilizer based on the application of spin freezing and IR-assisted drying, but this design might differ from the final, industrial configuration.

De Meyer et al. (2017) and Brouckaert et al. (2018) also investigated the use of Raman spectroscopy (Wikström et al. 2005) and NIR with the principal component analysis for the real-time monitoring of the residual moisture of the product being lyophilized during both primary and secondary drying. This methodology was proven to accurately detect the end point of each phase since the sublimation front is moving toward the glass wall and hence toward the sensor and, so, avoids the typical soak time that affects batch lyophilization. The application of thermal imaging completes the loop for adaptive control of the heat supply to each individual vial and hence maximizes drying efficiency without exceeding the critical product temperature (Van Bockstal et al. 2018b). This technology makes it possible to monitor and control the temperature at the sublimation front during primary drying in a noncontact way.

An important advantage of spin-freezing is that the frozen product is spread over the inner side wall of the vial, which makes it possible to dramatically increase the sublimation surface and decrease the layer thickness. This results in very high rates of sublimation and thus short drying times. The total drying duration was estimated to be reduced by a factor of 20 to 40 times. All measurement and heat transfer is executed on a noncontact basis. This means that there is no interference with the product during any of the process steps. This is an essential feature to ensure the quality of the product and the process of each individual vial, which is required for a continuous process and guarantees less vial-to-vial variability (one of the critical aspects that is inherent to conventional batch freeze-drying). However, some aspects still need to be further investigated. For example, some pharmaceutical ingredients

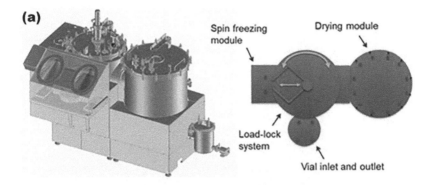

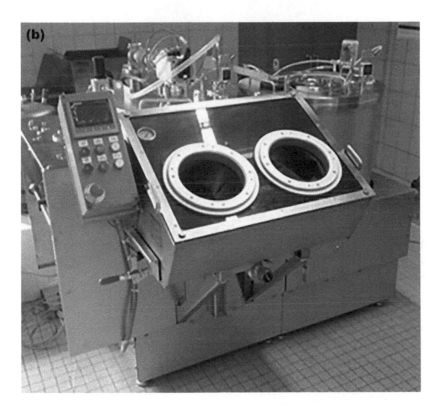

FIGURE 8.3   Schematic of the pilot-scale continuous freeze dryer (a). Picture of the pilot-scale continuous freeze-dryer (b) and the single-vial freeze-dryer (c). Reprinted with permission from RheaVita.

might be sensitive to the increase in air-to-liquid and air-to-glass area, and thus, appropriate changes to the formulation might be necessary.

At the date of writing, a pilot-scale prototype is used to execute the continuous process and a one-vial prototype has been realized (Corver 2017). Since the freezing and drying is controlled at the individual-vial level, the one-vial prototype process is

FIGURE 8.3 (Continued) Schematic of the pilot-scale continuous freeze dryer (a). Picture of the pilot-scale continuous freeze-dryer (b) and the single-vial freeze-dryer (c). Reprinted with permission from RheaVita.

identical to the continuous process in the pilot-scale prototype and future industrial production lines. Therefore, this one-vial prototype is a useful tool for process set-up and optimization and is also very useful in situations where very small amounts of drug material are available, such as in early stages of drug development.

8.5.3 CONTINUOUS LYOPHILIZATION IN UNIT DOSES, BASED ON CONVECTIVE FREEZING AND RADIATIVE DRYING OF A CONTINUOUS FLOW OF SUSPENDED VIALS

In 2015, the research group of Prof. Pisano at Politecnico di Torino and Prof. Trout at MIT proposed and filed a patent application for a new concept of continuous lyophilization of pharmaceuticals in the form of unit doses (Trout et al. 2018). The main idea behind this process is that a continuous flow of vials enters and leaves the apparatus, passing through different specialized chambers, as shown in Figure 8.4.

The process starts out from the continuous filling of vials, which, at that point, are suspended over a moving track and moved into the conditioning module. Here,

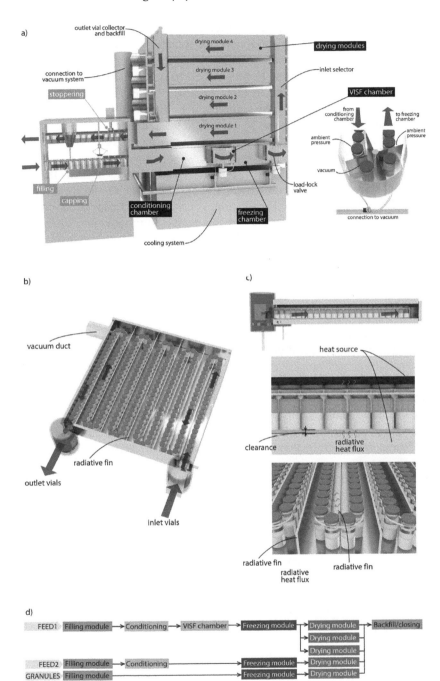

FIGURE 8.4 Schematic of (a) the continuous freeze-dryer and (b) a drying module. The front view of the drying module (c) and examples of potential configurations for the continuous lyophilizer (d) are also shown. The estimated chamber volume of the continuous lyophilizer (including both freezing and drying modules) is approximately 0.7 m³. Reprinted with permission from Capozzi et al. (2019c).

Source: Copyright (2019) American Chemical Society.

the vials are equilibrated at the desired temperature by forced convection with a cryogenic gas. Then, the precooled vials are transferred into a special chamber, the nucleation chamber, where the pressure is low enough to induce nucleation via vacuum-induced surface freezing (Oddone et al. 2014, 2015, 2016, 2017). Nucleated vials are then transferred to the freezing module, where, again, a cryogenic gas cools down the vial, achieving the complete solidification of the product. In this configuration, the freezing rate can be controlled by adjusting either temperature or flow rate of the cryogenic gas. Frozen vials are finally moved on to the drying module through a vacuum pass-through connector, which allows the transfer of the vials from chambers operating at different pressures and temperatures. In the drying module, vials are suspended over a track and move through the following module, for example, a snake-type path. The module is constituted of temperature-controlled walls that supply heat to the product via low-temperature radiation. Changing the wall temperature, it is possible to modulate the heat flow transferred to the product and, hence, to carry out both gentle and aggressive cycles. Once the vials are lyophilized, they are transferred to the backfilling and vial stoppering section. The entire process is carried out continuously, without breaks between phases or manual intervention.

One of the advantages of this technology is the control of the product structure, which can be achieved by the application of VISF and convective freezing. VISF promotes the nucleation of all the samples at the desired temperature, avoiding any variations in freezing history of the product due to the stochastic nature of spontaneous nucleation. Furthermore, the control of nucleation temperature and of cooling rate can in principle be used to control the morphological attributes of the finished products (Pisano and Capozzi 2017). In recent years, VISF has become increasingly popular among researchers, and a wide literature already exists (Fang et al. 2017; Pisano et al. 2017), making VISF a ready-to-use technique in continuous manufacturing.

A further advantage is the accurate control and uniformity of the heat supplied to the products during primary and secondary drying. Similar to the spin-freezing and IR-assisted drying, all the vials go through the same path and undergo identical drying conditions. Furthermore, contrary to the batch lyophilization, small variations in the geometry of the vials have no significant effect on the heat transferred to the vials, since it is essentially supplied by radiation. Finally, heat transfer by radiation is completely independent of chamber pressure, allowing further reduction of pressure during primary drying and, hence, increasing sublimation rate.

This method draws its strength from producing very uniform products, since vials undergo exactly the same processing conditions during both freezing and drying. It is extremely flexible because it can also treat particle-based materials in vials and it does not have a limitation in the shape of vessels. Capozzi et al. (2019c) showed that this process configuration reduces the primary drying duration by a factor of 2 to 4 times, and the total cycle time up to 10 times, including elimination of dead times. At this stage, a conceptual prototype is used to replicate similar conditions to those expected in the continuous configuration, while a pilot-scale prototype of the equipment is under construction.

8.6   CONCLUSION REMARKS

Table 8.1 summarizes the pros and cons of batch lyophilization and of the three continuous technologies described in section 8.5. Batch lyophilization is certainly a robust and established process, with an extensive literature and long industrial experience, although it still remains inefficient and expensive, with long dead time and limited throughput. Batch lyophilization has its weakness in the poor control of product quality uniformity, resulting in an unavoidable vial-to-vial and batch-to-batch variability of product characteristics.

TABLE 8.1
Comparison of Batch Freeze-Drying with Some Continuous Technologies

	Batch	AFD	RheaVita-Ghent University	Politecnico di Torino-MIT
Technology Status	Well established	Industrialization	Pilot-scale prototype; industrialization initiated	Conceptual prototype
Dosage Form	Bulk, Unit dose (liquid frozen solution, particle-based products), Cartridge	Particles	Unit doses (spin-frozen product), dual chamber cartridges, syringes	Unit doses (VISF-frozen products, particle-based products)
Fill Volume/Vial Dimension	Almost everything	N/A	Approx. 2/3 of the vial capacity	Almost everything
Dead Time	Very long	No	No	No
Processing Time	Very long	Shorter	Very short	Shorter
Equipment Footprint	Large	Smaller	Smaller	Smaller
Control of Product Structure	Possible	No	Yes, indirectly by control of the cooling and freezing cycle.	Yes, by control of nucleation
Scale-Up	Yes	Yes	Yes, duplication (numbering up)	Yes, duplication (numbering up)
Batch-to-batch variability	Yes	No	No	No
Vial-to-Vial Variability	Yes	N/A	No	No
In-Line Control	Lose	No	Yes	Yes
Process and Product Knowledge	Literature from multiple sources available	No	Literature from limited sources available	Literature from limited sources available

The continuous technology proposed by Hosokawa Micron BV does not produce end-to-use products, but only bulk materials in the form of fine particles. Handling this type of material (accurate filling of powder in containers is not easy), the lack of knowledge on the impact of this technology on product quality, and the inability to control product temperature and final moisture within the product might be, in our opinion, further weaknesses of this technology. However, this technology is close to being industrialized, and we expect to see new data and publications in the near future.

The two continuous technologies proposed for the lyophilization of unit doses have been designed to produce end-to-use products, avoiding the drawbacks of batch lyophilization in vials: processing time and equipment footprint are dramatically reduced, no manual operation or breaks are necessary, in-line control is naturally implemented, and scale-up simply consists of adding parallel modules. The use of spin-freezing is particularly advantageous in terms of extension of the area of sublimation and hence reduction of primary drying time. The use of VISF and convective freezing allows a precise control of freezing conditions (nucleation temperature and freezing rate), which is beneficial to the uniformity of the finished drug products. Overall, both technologies allow an increase in throughput and a decreased footprint of the manufacturing plant. Another advantage lies in the possibility of applying the technology to small batches as well. This facilitates the application of these technologies to early phases of new product development (e.g., for clinical trials). It is also applicable for specialized drugs, which require small quantities (personalized medicine). However, there are several barriers to the implementation of continuous lyophilization. For example, the cost of replacing traditional batch freeze-dryers with continuous apparatus is a significant investment. Therefore, it is debatable whether approved drugs that are produced with batch freeze dryers will be transferred to continuously operating freeze dryers. However, it is also true that this initial investment will be recovered over the years by the cost savings obtained, such as by the reduction of waste, reduced operational costs, reduction of batches lost owing to failure, and increased throughput. There is a stronger likelihood that initial implementation of continuous freeze-dryers is happening with new drug products that require new manufacturing environments. Appropriate real-time control and monitoring systems have to be developed to achieve well-controlled processes, similarly to what has been done in chemical plants. Furthermore, prior to the industrialization of these technologies, there is the need to adequately develop and engineer cleaning and sterilization strategies.

REFERENCES

Alexeenko, A. A., Ganguly, A., and S. L. Nail. 2009. Computational analysis of fluid dynamics in pharmaceutical freeze-drying. *J. Pharm. Sci.* 98:3483–3494.

Al-Fakih, E., Osman, N. A. A., and F. R. M Adikan. 2012. The use of fiber Bragg grating sensors in biomechanics and rehabilitation applications: The state-of-the-art and ongoing research topics. *Sensors* 12:12890–12926.

Allison, G., Cain, Y. T., Cooney, C., et al. 2015. Regulatory and quality considerations for continuous manufacturing. *J. Pharm. Sci.* 104:803–812.

Barresi, A. A., Pisano, R., Rasetto, V., Fissore, D., and D. L. Marchisio. 2010. Model-based monitoring and control of industrial freeze-drying processes: Effect of batch nonuniformity. *Drying Technol.* 28:577–590.

Baxendale, I. R., Braatz, R. D., Hodnett, B. K., et al. 2015. Achieving continuous manufacturing: Technologies and approaches for synthesis, workup, and isolation of drug substance. *J. Pharm. Sci.* 104:781–791.

Bosca, S., Fissore, D., and M. Demichela. 2015. Risk-based design of a freeze-drying cycle for pharmaceuticals. *Ind. Eng. Chem. Res.* 54:12928–12936.

Bosca, S., Barresi, A. A., and D. Fissore. 2017. On the robustness of the soft sensors used to monitor a vial freeze-drying process, *Drying Technol.* 35:1085–1097.

Breen, E. D., Curley, J. G., Overcashier, D. E., Hsu, C. C., and J. Shire. 2001. Effect of moisture on the stability of a lyophilized humanized monoclonal antibody formulation. *Pharm. Res.* 18:1345–1353.

Brouckaert, D., De Meyer, L., Vanbillemont, B., et al. 2018. The potential of near-infrared chemical imaging as process analytical technology tool for continuous freeze-drying. *Anal. Chem.* 90:4354–4362.

Bruttini, R. 1992. Continuous freeze-drying apparatus. U. S. Patent 5269077 A filed November 24, 1992.

Capozzi, L. C., and R. Pisano. 2018. Looking inside the 'black box': Freezing engineering to ensure the quality of freeze-dried biopharmaceuticals. *Eur. J. Pharm. Biopharm.* 129:58–65.

Capozzi, L. C., Barresi, A. A., and R. Pisano. 2019a. A multi-scale computational framework for modeling the freeze-drying of microparticles in packed-beds. *Powder Technol.* 343:834–846.

Capozzi, L. C., Barresi, A. A., and R. Pisano. 2019b. Supporting data and methods for the multi-scale modelling of freeze-drying of microparticles in packed-beds. *Data in Brief* 22:722–755.

Capozzi, L. C., Trout, B. L., and R. Pisano. 2019c. From batch to continuous: Freeze-drying of suspended vials for pharmaceuticals in unit-doses. *Ind. Eng. Chem. Res.* 58:1635–1649.

Chang, L., Shepherd, D., Sun, J., Tang, X., and M. J. Pikal. 2005. Effect of sorbitol and residual moisture on the stability of lyophilized antibodies: Implications for the mechanism of protein stabilization in the solid state. *J. Pharm. Sci.* 94:1445–1455.

Corver, J. A. W. M. 2013. Method and system for freeze-drying injectable compositions, in particular pharmaceutical compositions. W. O. Patent 2013036107 A2 filed August 27, 2012.

Corver, J. A. W. M. 2017. A continuous and controlled pharmaceutical freeze-drying technology for unit doses. Paper presented at the 8th International Conference of the International Society for Lyophilization and Freeze-Drying, La Habana, Cuba.

De Beer, T. R. M., Vercruysse, P., Burggraeve, A., et al. 2009. In-line and real-time process monitoring of a freeze drying process using Raman and NIR spectroscopy as complementary process analytical technology (PAT) tools. *J. Pharm. Sci.* 98:3430–3446.

Demarco, F. W., and E. Renzi. 2010. Bulk freeze-drying using spray freezing and stirred drying. U. S. Patent 9052138 B2 filed August 4, 2010.

De Meyer, L., Lammens, J., Mortier, S. T. F. C., et al. 2017. Modelling the primary drying step for the determination of the optimal dynamic heating pad temperature in a continuous pharmaceutical freeze-drying process for unit doses. *Int. J. Pharm.* 532:185–193.

De Meyer, L., Van Bockstal, P.-J., Corver, J., Vervaet, C., Remon, J. P., and T. De Beer. 2015. Evaluation of spin freezing versus conventional freezing as part of a continuous pharmaceutical freeze-drying concept for unit doses. *Int. J. Pharm.* 496:75–85.

Fang, R., Tanaka, K., Mudhivarthi, V., Bogner, R. H., and M. J. Pikal. 2017. Effect of controlled ice nucleation on stability of lactate dehydrogenase during freeze-drying. *J. Pharm. Sci.* 107:824–830.

Fissore, D., Pisano, R., and A. A. Barresi. 2019. Process analytical technology for monitoring pharmaceuticals freeze-drying – A comprehensive review. *Drying Technol.* 36:1839–1865.

Galan, M. 2016. Monitoring and control of industrial freeze-drying operations: The challenge of implementing Quality-by-Design (QbD). In *Freeze-drying/lyophilization of pharmaceutical and biological products*, ed. L. Rey and J. C. May, 453–471. New York: CRC Press.

Gasteyer, T. H., Sever, R. R., Hunek, B., Grinter, N., and M. L. Verdone. 2015. Lyophilization system and method, U. S. 9651305 B2 filed November 17, 2015.

Gieseler, H., and P. Stärtzel. 2012. Controlled nucleation in freeze-drying. *Eur. Pharm. Rev.* 17(5), 63–68.

Goldblith, S. A., Rey, L., and W. W. Rothmayr. 1975. *Freeze-drying and advanced food technology.* New York: Academic Press.

Goshima, H., Do, G., and K. Nakagawa. 2016. Impact of ice morphology on design space of pharmaceutical freeze-drying. *J. Pharm. Sci.* 105:1920–1933.

Keenan, J. 2014. Lyophilization market forecast to double in size by 2020. FiercePharma. www.fiercepharma.com/manufacturing/lyophilization-market-forecast-to-double-size-by-2020 (accessed March 31, 2019).

Lackner, M. 2007. Tunable diode laser absorption spectroscopy (TDLAS) in the process industries – A review. *Chem. Eng. Rev.* 23:65–147.

Lietta, E., Colucci, D., Distefano, G., and D. Fissore. 2019. On the use of infrared thermography for monitoring a vial freeze-drying process. *J. Pharm. Sci.* 108:391–398.

Lim, J. Y., Kim, N. A., Lim, D. G., Kim, K. H., Choi, D. U., and S. H. Jeong. 2016. Process cycle development of freeze drying for therapeutic proteins with stability evaluation. *J. Pharm. Investig.* 46:519–536.

Mihokovic, N. 2017. Continuous manufacturing – EMA perspective and experience. Paper presented at the Integrated Continuous Biomanufacturing III, Cascais, Portugal.

Myerson, A. S., Krumme, M., Nasr, M., Thomas, H., and R. D. Braatz. 2015. Control systems engineering in continuous pharmaceutical manufacturing. *J. Pharm. Sci.* 104:832–839.

Nasr, M. M., Krumme, M., Matsuda, Y., and B. L. Trout. 2017. Regulatory perspectives on continuous pharmaceutical manufacturing: Moving from theory to practice. *J. Pharm. Sci.* 106:3199–3206.

Oddone, I., Barresi, A. A., and R. Pisano. 2017. Influence of controlled ice nucleation on the freeze-drying of pharmaceutical products: The secondary drying step. *Int. J. Pharm.* 524:134–140.

Oddone, I., Fulginiti, D., Barresi, A. A., Grassini, S., and R. Pisano. 2015. Non-invasive temperature monitoring in freeze-drying: Control of freezing as a case study. *Drying Technol.* 33:1621–1630.

Oddone, I., Pisano, R., Bullich, R., and P. Stewart. 2014. Vacuum-Induced nucleation as a method for freeze-drying cycle optimization. *Ind. Eng. Chem. Res.* 53:18236–18244.

Oddone, I., Van Bockstal, P.-J., De Beer, T., and R. Pisano. 2016. Impact of vacuum-induced surface freezing on inter-and intra-vial heterogeneity. *Eur. J. Pharm. Biopharm.* 103:167–178.

Oetjen, G. W. 1999. Industrial freeze-drying for pharmaceutical applications. In *Freeze-drying/lyophilization of pharmaceutical and biological Product*, ed. L. Rey and J. C. May, 421–471. New York: Informa Healthcare.

Oetjen, G. W., Schmitz, F. J., and H. Eilenberg. 1969. Continuous freeze dryer, U. S. Patent 3612411 A filed August 6, 1969.

Patel, S. M., Nail, S. L., Pikal, M. J., et al. 2017. Lyophilized drug product cake appearance: What is acceptable? *J. Pharm. Sci.* 106:1706–1721.

Peters, B.-H., Staels, L., Rantanen, J., et al. 2016. Effects of cooling rate in microscale and pilot scale freeze-drying. Variations in excipient polymorphs and protein secondary structure. *Eur. J. Pharm. Sci.* 95:72–81.

Pikal, M. J., and S. Shah. 1997. Intra-vial distribution of moisture during the secondary drying stage of freeze drying. *PDA J. Pharm. Sci. Technol.* 51:17–24.

Pikal, M. J., Bogner, R., Mudhivarthi, V., Sharma, P., and P. Sane. 2016. Freeze-drying process development and scale-up: Scale-up of edge vial versus center vial heat transfer coefficients, K_v. *J. Pharm. Sci.* 105:3333–3343.

Pikal, M. J., Roy, M. L., and S. Shah. 1984. Mass and heat transfer in vial freeze-drying of pharmaceuticals: Role of the vial. *J. Pharm. Sci.* 73:1224–1237.

Pisano, R., et al., 2013a. Quality by Design: Scale-up of freeze-drying cycles in pharmaceutical industry. *AAPS PharmSciTech* 14:1137–1149.

Pisano, R., et al., 2013b. Quality by design: Optimization of a freeze-drying cycle via design space in case of heterogeneous drying behavior and influence of the freezing protocol. *Pharm. Dev. Technol.* 18:280–295.

Pisano, R., et al., 2015. Effect of electron beam irradiation on remaining activity of lyophilized acid phosphatase with water-binding and non-water – binding additives. *Drying Technol.* 33:822–830.

Pisano, R., et al., 2017. Characterization of the mass transfer of lyophilized products based on X-ray micro-computed tomography images. *Drying Technol.* 35:933–938.

Pisano, R. et al., 2017. Prediction of product morphology of lyophilized drugs in the case of Vacuum Induced Surface Freezing. *Chem. Eng. Res. Des* 125:119–129.

Pisano, R. et al., 2019. Achieving continuous manufacturing in lyophilization: Technologies and approaches. *Eur. J. Pharm. Biopharm.* 142: 265–279.

Pisano, R. 2019. Alternative methods of controlling nucleation in freeze-drying. In *Lyophilization of pharmaceuticals and biologicals*, ed. K. R. Ward and P. Matejtschuk, 79–111. New York: Humana Press.

Poechlauer, P., Manley, J., Broxterman, R., Gregertsen, B., and M. Ridemark. 2012. Continuous processing in the manufacture of active pharmaceutical ingredients and finished dosage forms: An industry perspective. *Org. Proc. Res. Dev.* 16:1586–1590.

Rambhatla, S., Ramot, R., Bhugra, C., and M. J. Pikal. 2004. Heat and mass transfer scale-up issues during freeze drying: II. Control and characterization of the degree of supercooling. *AAPS PharmSciTech* 5:54–62.

Rasetto, V., Marchisio, D. L., Fissore, D., and A. A. Barresi. 2010. On the use of a dual-scale model to improve understanding of a pharmaceutical freeze-drying process. *J. Pharm. Sci.* 99:4337–4350.

Rey, L. 2010. Glimpses into the realm of freeze-drying: Classical issues and new ventures. In *Freeze-drying/lyophilization of pharmaceutical and biological products*, ed. L. Rey and J. C. May, 1–32. New York: CRC Press.

Searles, J. A., Carpenter, J. F., and T. W. Randolph. 2001. The ice nucleation temperature determines the primary drying rate of lyophilization for samples frozen on a temperature-controlled shelf. *J. Pharm. Sci.* 90:860–871.

Scutellà, B., Passot, S., Bourlés, E., Fonseca, F., and I. C. Tréléa. 2017. How vial geometry variability influences heat transfer and product temperature during freeze-drying. *J. Pharm. Sci.* 106:770–778.

Sluder, J. C., Olsen, R. W., and E. M. Kenyon. 1947. A method for the production of dry powdered orange juice. *Food Technol.* 1:85–94.

Touzet, A., Pfefferlé, F., van der Wel, P., Lamprecht, A., and Y. Pellequer. 2018. Active freeze drying for production of nanocrystal-based powder: A pilot study. *Int. J. Pharm.* 536:222–230.

Trout, B. L., Pisano, R., and L. C. Capozzi. 2018. Continuous freeze-drying methods and related products, International Patent Application WO 2018/204484 A1 filed May 2, 2018.

Van Bockstal, P. J., De Meyer, L., Corver, J., Vervaet, C., and T. De Beer. 2017a. Noncontact infrared-mediated heat transfer during continuous freeze-drying of unit doses. *J. Pharm. Sci.* 106:71–82.

Van Bockstal, P. J., Mortier, S. T. F. C., Corver, J., Nopens, I., Gernaey, K. V., and T. De Beer. 2017b. Quantitative risk assessment via uncertainty analysis in combination with error propagation for the determination of the dynamic Design Space of the primary drying step during freeze-drying. *Eur. J. Pharm. Biopharm.* 121:32–41.

Van Bockstal, P. J., Mortier, S. T. F. C., De Meyer, L., et al. 2017c. Mechanistic modelling of infrared mediated energy transfer during the primary drying step of a continuous freeze-drying process. *Eur. J. Pharm. Biopharm.* 114:11–21.

Van Bockstal, P. J., Corver, J., De Meyer, L., Vervaet, C., and T. De Beer. 2018a. Thermal imaging as a noncontact inline process analytical tool for product temperature monitoring during continuous freeze-drying of unit doses. *Anal. Chem.* 90:13591–13599.

Van Bockstal, P. J., Corver, J., Mortier, S. T. F. C., et al. 2018b. Developing a framework to model the primary drying step of a continuous freeze-drying process based on infrared radiation. *Eur. J. Pharm. Biopharm.* 127:159–170.

van der Wel, P. G. J. 2004. Stirred freeze drying, E. U. Patent 1601919 A2 filed February 13, 2004.

van der Wel, P. G. J. 2015. Active freeze drying. Paper presented at the 5th European Drying Conference, Budapest, Hungary.

van der Wel, P. G. J. 2017. Active freeze drying. Paper presented at the 8th International Conference of The International Society for Lyophilization and Freeze-Drying, La Habana, Cuba.

Velardi, S. A., Hammouri, H., and A. A. Barresi. 2010. Development of a high gain observer for in-line monitoring of sublimation in vial freeze drying. *Drying Technol.* 28:256–268.

Vollrath, I., Pauli, V., Friess, W., Freitag, A., Hawe, A., and G. Winter. 2017. Evaluation of heat flux measurement as a new process analytical technology monitoring tool in freeze drying. *J. Pharm. Sci.* 106:1249–1257.

Wang, B., McCoy, T. R., Pikal, M. J., and D. Varshney. 2015. Lyophilization of therapeutic proteins in vials: Process scale-up and advances in quality by design. In *Lyophilized biologics and vaccines*, ed. D. Varshney and M. Singh, 121–156. New York: Springer.

Wikström, H., Lewis, I. R., and L. S. Taylor. 2005. Comparison of sampling techniques for in-line monitoring using Raman spectroscopy. *Appl. Spectrosc.* 59:934–941.

9 Use of CFD for the Design and Optimization of Freeze-Dryers

Antonello Barresi

CONTENTS

9.1 INTRODUCTION

In freeze-drying, the design of the equipment (chamber, duct, valve, and condenser) has a strong impact on the final product quality, on the flexibility and reliability of the process, and on its optimal operation and duration. In fact, for example, the required minimum pressure in the chamber can be obtained with an effective pressure control, together with the duct connecting it to the chamber and the isolating valve, only if the condenser was correctly designed and properly sized.

When transferring and especially when scaling-up the process from one equipment to a different one, it is generally necessary to adapt the cycle; but it must also be verified that the equipment geometry will allow the expected maximum sublimation rate with the required intrabatch variance.

Those mentioned above are critical issues that must be addressed when designing or evaluating a freeze-dryer.

The piece of equipment must be able to evacuate the requested water vapor flow rate at the desired operating conditions (i.e., specific sublimation rates and thus batch drying times and chamber pressure). Problems may arise in production at the industrial scale, when the equipment is operating at full load, differently from what generally happens during the development stage; the sublimation surface, and thus the vapor mass flow rate, generally increases more in the industrial scale, where smaller clearances between shelves are also allowed, that in the the the duct.

The intrabatch variance is related to the problem of batch heterogeneity; in fact, the product contained in vials positioned in different points of the chamber experience different temperature and pressure histories, resulting in significant variations for the drying time and the final residual water content. The relevant design parameters are chamber geometry, clearance between shelves, number of shelves, position of the duct leading to the condenser, number and position of the inert gas injection nozzles, as well as temperature gradients of the heating fluid circulating through the shelves (Barresi et al. 2010b).

9.1.1 USE OF CFD

Since the experimental investigation of these issues is costly and time-consuming, the use of a computational approach to obtain meaningful predictions is very interesting.

When the Knudsen number is much smaller than one, the fluid can be considered as a continuum and the standard continuity (mass conservation), momentum, and energy balance equations can be solved by using standard computational fluid dynamics tools, based on the solution of the governing equations with finite-volume schemes with no-slip boundary conditions at the walls (Batchelor 1965). Under usual operating conditions (both for chamber and condenser) and for the most common industrial geometries this condition is almost always satisfied, although at the low pressures typical of freeze-drying operations, the characteristic mean free path of the involved gas molecules can be quite large. Inside valves, or with very small shelf-clearances, the transitional regime may occur ($0.01 < Kn < 0.1$); in this case CFD equations are solved with slip boundary conditions at the walls owing to the small number of collisions with the walls (Maxwell 1879)

This approach has been successfully employed to model the drying chamber, to investigate the internal fluid dynamics, to predict possible nonuniformity of the batch, to analyze critical aspects in scale up and process monitoring (Rasetto 2009; Barresi et al. 2010a, 2010b), to predict the spatial variation of pressure and inert fraction over the shelf (Rasetto et al. 2010; Ganguly et al. 2017; Barresi et al. 2018), to model flow in the cylindrical duct, and to estimate conductance and calibrate the TDLAS measuring system (Alexeenko et al. 2009; Patel et al. 2010; Marchisio et al. 2018).

CFD can be used to simulate pseudostationary conditions or the entire freeze-drying cycle by adopting a multiscale approach, which allows consideration of mutual interactions between chamber hydrodynamics and product sublimation rate (Rasetto at al. 2010; Barresi et al. 2010a; Marchisio et al. 2018). The two-scale CFD model can be also used for design purposes: by fixing the maximum allowed sublimation rate difference, in fact, it is possible to calculate the maximum pressure difference across a single shelf or between different shelves, or to calculate the maximum tolerated gradients in the heating fluid circulating inside the shelf.

The analysis can be limited to specific parts of the equipment (i.e., chamber, condenser, duct, valve, etc.) or can include the entire apparatus. CFD has been used to test different design solutions, for example, to show the effect of location of the duct connecting the chamber to the condenser, to evaluate the effect of equipment modifications, valve type, shape, and positioning (Ganguly et al. 2013; Marchisio et al. 2018; Barresi and Marchisio 2018; Zhu et al. 2018; Kshirsagar et al. 2019).

Under moderate operating conditions, the Mach number is smaller than unity and the flow is described with incompressible CFD solvers. Anyway, sonic flow conditions may typically occur in the duct, in correspondence with the reduced section of the valve, in the case of choked flow, and may occur also in the chamber, in the case of jet flow caused by leakage from small holes. In this case it is necessary to employ compressible CFD solvers capable of dealing with sonic and super-sonic conditions and with shockwaves. Recently it was demonstrated that at high sublimation rates the vapor flow can be characterized by turbulent effects, which cause an increase of the minimum controllable pressure in the chamber (Kshiragar et al. 2019)

At very low pressure, when the Kn number is larger than unity, the continuum hypothesis is not valid anymore. Under these conditions the flow must be described by molecular dynamics simulations or by solving the Boltzmann equation. These conditions may typically occur inside the condenser.

CFD can still be employed in many cases, anyway, and, for example, has been used to estimate the flow field in different types of condensers and around a mushroom valve (Petitti et al. 2013) or the effect of a baffle (Ganguly et al. 2013) at the entrance of a condenser. Computations can become extremely heavy in this case, especially for the complex geometries of industrial apparatus, owing to the necessity of modeling the vapor disappearance (and the ice formation) with a realistic mechanism that considers the proper kinetics.

At lower pressure conditions it may be necessary to adopt other modeling approaches more suitable for the transitional and free-molecular regimes, like the direct simulation Monte Carlo (DSMC) as proposed by Ganguly and Alexeenko (2012); this method was also very effective in describing the icing phenomena that occurs on the coils of the condenser.

A detailed account of the different modeling approaches and equations, together with a review of previous work, can be found in recent articles (Barresi et al. 2018; Marchisio et al. 2018); sets of CFD data usable for design are also available (Barresi and Marchisio 2018). Use of CFD as a PAT tool to achieve quality-by-design has been also presented in previous book chapters (Barresi et al. 2010a; Fissore et al. 2015). The aim of this chapter is to present and discuss a procedure to design or verify performances of a freeze-dryer, using data from CFD simulations.

The complete design procedure, for a given chamber geometry, can include the steps listed in Table 9.1. The correlations and the data presented will allow a good design procedure for general equipment independent of its geometry. Of course, some of the steps (see in particular b.2, c, and partly d in Table 9.1) require additional information, obtainable only from a detailed analysis of the considered equipment through CFD. The work by Patel et al. (2010) demonstrated that a correlation developed on one freeze-dryer cannot be easily generalized and applied to other

TABLE 9.1
Outline of a General Design Procedure for the Chamber

a) **Selection of the nominal chamber pressure for maximum expected sublimation flow rate and preliminary estimation of the duct conductance**

b) **Evaluation of pressure conditions over shelves**

> **b.1)** estimation of the pressure drop along a nominal shelf (neglecting duct position effects)
>
> **b.2)** estimation of the real pressure distribution over the different shelves
>
> *If the estimated pressure variation is not negligible*:
>
> > **b.1.1)** estimation of the intravial variance due to pressure gradients over the standard shelf
> >
> > **b.2.1)** estimation of the real intravial variance due to pressure distribution over the different shelves of the freeze-dryer

(steps b.2 and b.2.1 require that the pressure profiles over the different shelves, depending on the relative position with respect to the duct, have been preliminary estimated through CFD)

> *If the pressure variation estimated at step b.1 and eventually b.2 (or the intravial variances estimated with the more complex procedure at steps b.1.1 or b.2.1) exceeds the required limits, the clearance between the shelves must be increased. (*)*

c) **Estimation of the chamber resistance and pressure drop caused by the duct entrance effect**

> **c.1)** estimation of pressure drop from nominal measuring point to duct entrance
>
> **c.2)** estimation of pressure drop due to entrance effects and of actual entrance pressure in duct

d) **Verification of choked flow conditions and minimum allowable chamber pressure**

> **d.1)** Using the actual inlet pressure (see c.2), estimate maximum mass flux allowed by the chamber pressure set, including the effect of the inert gas in the flow from the chamber on the duct conductance for the considered duct with given L/D or for the considered valve (**)
>
> *If allowed maximum flux is lower than required (choked flow), alternatively*:
>
> > d.1.1 calculate required duct and valve size
> >
> > d.1.2 calculate chamber pressure required for the given mass flow rate and duct size
>
> **d.2)** Using design charts, calculate the expected pressure drop and minimum controllable chamber pressure for the considered design and operating conditions (verify pressure selected in a)

(*) Please note that the clearance considered by the proposed correlation is the **free space between the top stopper plane and the upper shelf;** thus, it depends not only on the actual shelf clearance but also on the vial size used by the customer.
(**) It must be noted that the butterfly valve is generally responsible for the largest part of the resistance in ducts of practical length; the use of an equivalent length duct to consider the presence of the valve can be accepted but gives approximated results.

freeze-dryers, and some details of the apparatus geometry can have a significant effect on hydrodynamic, as shown in Barresi and Marchisio (2018).

In order to illustrate the procedure and highlight the advantage of using CFD for design and optimization, some typical results obtained from the simulation of a real industrial apparatus equipped with a 10 m³ chamber and either butterfly or mushroom isolating valve will be presented.

In the industrial chamber the number of usable shelves has been varied from 14 to 17, but the position of the first and of the last shelf was not varied. The distance between the shelves varies from 110 to 85 mm with no load, but a product thickness of 43 mm has always been considered, which reduces the space available to vapor flow accordingly. The largest clearance corresponds to normal and usual conditions in practical cases, while narrower clearances may represent cases where larger loading is obtained, increasing the number of shelves, or taller containers are used.

9.2 CHAMBER PRESSURE AND ESTIMATION OF THE DUCT CONDUCTANCE

Operating chamber pressure and/or sublimation flow rate are generally given as process specifications and may be the results of a process optimization and of a risk evaluation.

It must be taken into account that the drying rate increases if the operation is carried out at higher temperature and higher chamber pressure, because higher pressure increases the heat transfer rate from shelf to product in vials or trays (Pisano et al. 2011); this effect is often prevailing on the reduction of the driving force for mass transfer, especially in the first part of the primary drying. On the other hand, to operate at higher pressure and temperature increases the risk of trespassing the limit product temperature, thus causing the failure of the processed batch.

At high sublimation rates, choked flow conditions can occur if the equipment is not properly designed (Searles 2004; Patel et al. 2010).

A preliminary check for the consistency of the data can be easily done by using simplified jet-flow calculations or choked-flow design charts for ducts, which allow correlation of maximum mass flux (that is, the mass flow rate in the duct divided by the duct cross section) with the chamber pressure; from the value of the mass flux, the required nominal diameter of duct and valve can be easily calculated. In case of valves, an equivalent L/D ratio can be assumed, in order to use the chart valid for the ducts: this is not very accurate, as will be shown in the following text, but it is acceptable for this preliminary calculation.

The maximum mass flow rate, $\dot{m}_{max}$, in the case of "jet flow" can be calculated using the following equation, where k is the specific heat ratio ($k = c_p / c_v$), M is the molecular weight, ρ^* and u^* are the density and velocity of the fluid, and A^* is the restricted section when the flow is sonic (White 2009):

$$\dot{m}_{max} = \rho^* A^* u^* = P_0 \left(\frac{2}{k+1} \right)^{\frac{1}{(k-1)}} A^* \left(\frac{2k}{k+1} \frac{M}{RT_0} \right)^{\frac{1}{2}}. \qquad (9.1)$$

T_0 and P_0 appearing in the formula are the stagnation temperature and pressure, which correspond to inlet temperature and pressure only in the case where the initial gas velocity is null (or at least very low, compared to the sonic one). For duct flow the velocity at the inlet is of the same order of magnitude as that at the exit and should be considered, to reduce the estimation error. It is in any case a simplified calculation, in which only the behavior along the centerline is considered, not taking into account the effect of the wall, and assuming isentropic behavior; a comparison with exact solutions is shown in Barresi and Marchisio (2018) [Figures 2 and 21].

9.2.1 Design Charts

More accurate estimations can be carried out using duct design charts, of which different examples are shown in Figure 9.1.

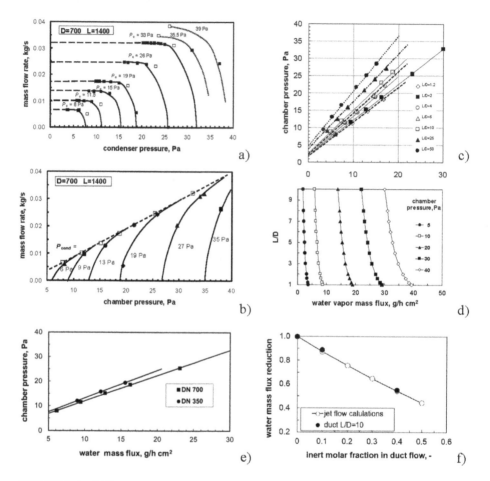

FIGURE 9.1 Different design charts for duct conductance. Upper left: (a) Water mass flow rate as a function of the condenser pressure for different chamber pressures, P_c. The transition to sonic flow condition, with constant mass flow rate, is indicated by the dashed line. (b) Water mass flow rate as a function of the chamber pressure for different condenser pressures, P_{cond}. The dashed line with open symbols represents the asymptote and corresponds to sonic flow. Straight duct, DN 700, $L = 1400$ mm ($L/D = 2$).

Upper right: Relationship between maximum water mass flux in straight ducts of different L/D ratio and chamber pressure (c) or L/D ratio (d). The simulations have been carried out for DN 700, but the results have been generalized using the mass flux concept.

Bottom graphs: (e) Relationship between critical water mass flux and chamber pressure for different duct diameters ($L/D = 2$). (f) Reduction in the critical water mass flux as a function of the inert gas fraction in mixture for jet flow and in a long duct (the actual value of the flux divided by the value with no inert gas is plotted).

Plots (a–b) and (d–f) reprinted from Marchisio et al. (2018), with permission from Elsevier. Plot (c), from Barresi and Marchisio (2018).

The charts have been obtained by detailed CFD simulations carried out for ducts with DN 700; plots (a) and (b) show the points obtained from CFD simulation, and the fitting curve, which allows calculation of mass flow as a function of the chamber (inlet) and condenser (outlet) pressure. Pressure is not uniform on the duct section, and attention must be paid to the consistent use of pressure definition and correlations: those shown are relative to the pressure average over the section.

In Figure 9.1a it can be noted that mass flow rate always increases with chamber pressure, while reducing the condenser pressure increases the mass flow rate up to a limit value, after which remains constant: these are the choked flow conditions. Figure 9.1b is an alternative representation of the same data, which demonstrates the asymptotic values obtained when sonic flow occurs. This type of charts gives a lot of information, but they are specific to a certain L/D value and depend on the duct diameter; another example for a smaller diameter has been shown elsewhere (Fissore et al. 2015). The graphs can be generalized by plotting the mass flux (that is, the mass flow rate divided by the section area), but for ducts with different diameter, and in particular for smaller ones, a diameter correction factor should be used; the correction required is anyway relatively small, as shown in Figure 9.1e, but a small safety factor must be introduced when using smaller sizes, as the conductance of smaller ducts is slightly lower.

Considering only the data for which clearly sonic conditions had been obtained, information for ducts of different lengths can be synthesized in a single plot. Figures 9.1c–d show an example of design charts for choked flow conditions. The first one highlights the relationship between chamber pressure and mass flux, while the second one shows the influence of the L/D ratio on mass flux, at different chamber pressures. The range 1–5 is the one of largest interest, and data available in the literature were limited to this range, but here the range from 1.2 to 50 has been considered to include values that can correspond to equivalent length of valves (from 10 to 40 typically).

These results can be compared with those published by Oetjen (1999) and calculated with the Günter-Jaeckel-Oetjen equation (the units of the mass flux have been selected to allow an easy comparison with previous literature data and to give values in a comfortable range). The condenser pressure is not considered when the maximum flux is plotted, as it is assumed that it is lower than the value that determines choked flow conditions.

It is evident that a linear relationship exists between the inlet (chamber) pressure and the critical mass flux. The lines shown in the graph have been calculated by combined fitting of the whole data set obtained by simulation of DN 700 ducts.

The chamber (inlet) pressure *vs.* critical mass flux ($j_{w,c}$) (g/h cm²) relationship is described by Equation 9.2, which has been used to draw lines in the showed plots. The constants depend on the L/D ratio and on the duct diameter; two parameters have been introduced to consider the effect of the two variables separately:

— f_{DN} is a correction factor for the effect of the duct diameter and is taken equal to 1 for DN 700;
— α_{P0} is given by Equation 9.3, as all the curves pass for the same point.

$$P_{in} = f_{DN}\,\alpha_P\,j_{w,c} + \alpha_{P0} \tag{9.2}$$

$$\alpha_{P0} = 5f_{DN}\alpha_P - 3.5 \tag{9.3}$$

The parameter α_P depends on the L/D ratio: for $L/D = 2$, $\alpha_P = 1.048$; two families can be recognized, with different correlations for shorter and longer ducts:

$$L/D < 10 \qquad \alpha_P = 0.98\left(\frac{L}{D}\right)^{0.1}, \tag{9.4}$$

$$10 < L/D < 50 \qquad \alpha_P = 0.009\left(\frac{L}{D}\right) + 1.15, \tag{9.5}$$

Mass flux can be related to the pressure drop and to the inlet pressure also in the subcritical region. Correlations and plots available in the technical literature deal generally with developed flow or assume stagnant fluid at the inlet, but in real cases, especially in the short ducts that are those of major interest, the velocity profile is still developing.

The dependence is relatively complex not only because the compressibility of the gas must be considered but also because in subcritical flow the dependence on the duct size is important (at the critical flow conditions the duct diameter had a very weak influence). Anyway, it is still possible to propose an empirical correlation between pressure drop (ΔP) (Pa), inlet (or chamber) pressure, and mass flux (j_w) (g/h cm²):

$$\Delta P\, P_{in}^{1.5} = g_{DN}\, m\, j_w^{exp}, \tag{9.6}$$

where
— g_{DN} is a correction factor for the effect of the duct diameter and is taken equal to 1 for DN 700,
— "m" and "exp" are two parameters introduced to consider the effect of L/D and duct diameter separately: for $L/D = 2$, m = 0.98, and exp = 2.07.

The small variation in the exponent, which remains anyway very close to 2, can be justified considering that the pressure drop is mainly due to the development of the velocity profile.

$$m = 0.28\left(\frac{L}{D}\right) + 0.42, \tag{9.7}$$

$$exp = -0.01\left(\frac{L}{D}\right) + 2.09. \tag{9.8}$$

For the subcritical flow correlation given above, only those cases where the local Mach number was below 0.9 have been considered. For higher velocity values, which were considered in the "transonic" region, an analysis carried out in detail for the butterfly valve showed that the mass flux is mainly dependent on the inlet (chamber) pressure but has also a weak dependence on the outlet pressure. These results have been verified to be applicable also to duct flow, confirming that in transonic condition, for each L/D set, data can be correlated considering that the mass flux is proportional to the chamber pressure, P_c, corrected by the chamber/condenser pressure ratio, P_c/P_{cond} (an exponent equal to 0.2 gave satisfactory results also for ducts).

Thus, a correlation with the structure given in Equation 9.9 can be used with good results:

$$j_w = \beta_0 + \beta P_c \left(\frac{P_c}{P_{cond}} \right)^{0.2}. \qquad (9.9)$$

By using the design equations given above, Equations 9.7–9.8 for subcritical flow and 9.2–9.5 for choked flow, it is possible to realize design charts that show immediately the dependence of mass flux on chamber (and condenser) pressure for the duct with desired size.

In presence of bends and elbows, the previous calculations for a straight duct can be applied provided an equivalent length is assumed for the line. It has been proposed (*Leybold Vacuum Products and Reference Book*, 2001/2) that where the line contains elbows or other curves (such as in right-angle valves), they can be considered by assuming a greater effective length of the line. This can be estimated by adding to the axial length of the line 1.33 $(\theta/180°)$ D, where θ is the angle of the elbow (degrees of angle). This should be regarded as an approximated approach, as the presence of bends generates a more complicated flow characterized by secondary recirculation.

9.2.2 EFFECT OF INERT GAS ON DUCT CONDUCTANCE

The presence of an inert gas in mixture with the water vapor in significant amounts can modify the conductance of the duct or valve.

For design purposes it is important to evaluate which is the reduction of the water mass flux as a function of the inert fraction in the mixture. The simple correlation (9.10), valid for both jet flow and flow in duct, has been obtained and can be reasonably used also for valves:

$$\frac{j_{mix}}{j_w} = 1 - 1.13 y_{N_2}. \qquad (9.10)$$

It gives the relative values of the water mass flux of the mixture (j_{mix}) with respect to that with pure water vapor (j_w) as a function of the nitrogen molar fraction, y_{N2}. Figure 9.1f shows that the exact solution obtained by CFD in a long duct corresponds, in terms of relative variation, to the simple estimation obtained by jet flow calculations.

9.3 EVALUATION OF PRESSURE CONDITIONS OVER SHELVES

The position of the duct strongly influences the pressure distribution over the shelves: a configuration with the duct positioned in the bottom of the chamber would generate a more uniform situation over the shelves and a symmetric pressure profile (Barresi et al. 2018; Kshiragar et al. 2019).

The variation of pressure over the shelf in any case has relevant effects only with large shelf size, small clearance, and at high sublimation rate; the increase in the clearance between the shelves eliminates any problem but obviously reduces the apparatus loading.

Figures 9.2a–b show an example of the pressure profile over shelves in different positions in the freeze-dryer, demonstrating the symmetry of those far from the duct

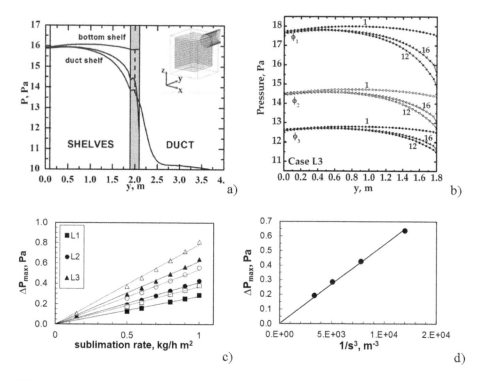

FIGURE 9.2   (a) Pressure gradients inside an industrial freeze-dryer; profiles in the duct and in the drying chamber for three shelves (the shelf at the bottom of the chamber, close to the duct, and at the top of the chamber) are shown (the grey zone corresponds to the clearance between wall and shelves). L1 configuration, 14 shelves, $s = 67$ mm; $J_w = 1$ kg h^{-1}m^{-2}.

(b) Pressure profiles over some shelves in the large-scale apparatus (L3 configuration, 16 shelves, $s = 50.5$ mm) along the mean x position; the numbers identify the shelf, starting from the bottom. Values obtained at different mass sublimation flux: 1 kg h^{-1}m^{-2} (case $\Phi1$), 0.7 kg h^{-1}m^{-2} (case $\Phi2$), and 0.5 kg h^{-1}m^{-2} (case $\Phi3$).

(c) Maximum overpressure (filled symbols) and maximum pressure difference (open symbols) over the bottom shelf of the industrial freeze-dryer as a function of the sublimation rate, for different shelf configurations.

(d) Maximum overpressure as a function of the actual clearance for the same system. $J_w = 1$ kg h^{-1}m^{-2}.

Plot (a) reprinted from Barresi et al. (2018), with permission from Elsevier. Plot (b) reprinted from Barresi et al. (2010a) by permission of the publisher (Taylor & Francis Ltd, www.tandfonline.com).

(a lateral upper duct is considered for a large-scale industrial apparatus), and the increase of pressure variation along the shelf at higher sublimation rate; in this case a constant pressure has been considered at the duct exit in the different cases, thus Figure 9.2b also shows the effect of sublimation rate on minimum pressure achievable in the chamber, if pressure drop in the camber-duct section is not negligible.

9.3.1 ESTIMATION OF THE PRESSURE DROP ALONG A NOMINAL SHELF

For shelves not significantly influenced by the duct (in the case of the duct on the bottom of the chamber, or for shelves far from the duct entrance) the CFD simulations have confirmed the validity of correlations developed from a theorical approach. In Barresi et al. (2010b) the dependence of the pressure drop on the different variables had been shown and validated, demonstrating how correlations can be used, for example, to develop a dual scale model. The general correlation is given in (9.11), and is shown in Figure 9.3 (dash-dotted line), where the symbols refer to CFD results for pressure profiles calculated on the bottom shelf of the industrial freeze-dryer with different chamber pressure, sublimation flux, and clearance:

$$\Delta P = \beta \frac{J_w L_s^2}{P_{ref} s^3}, \qquad (9.11)$$

where ΔP_{max} is the maximum pressure variation along the shelf (Pa); J_w is the sublimation flux on the shelf (kg/h m²); L_s is the shelf distance, that is, the length from the

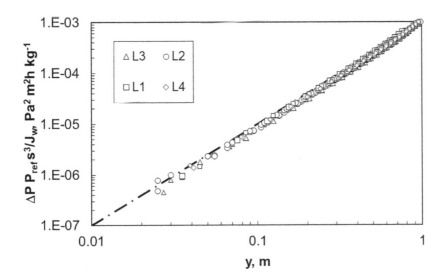

FIGURE 9.3 Correlation of pressure drop along a shelf, in the y direction, with limited asymmetry, validated with CFD simulations (over different shelves, considering the distance y from the maximum pressure position). Configurations: L1, 14 shelves; L2, 15 shelves; L3, 16 shelves; L4, 17 shelves (actual clearance, s, variable from 44 to 67 mm). J_w variable from 0.15 to 1.0 kg h⁻¹m⁻².

location of the maximum to the shelf border (m); L_{max} can be half the shelf size if the duct is not influencing the profile, and it can become closer to the shelf size for the part of the shelf close to the duct; P_{ref} is the "reference pressure" in the chamber (Pa) (the pressure in the zone far from the duct is considered and should correspond to the pressure "measured" in the chamber); s is the actual clearance between the shelves, that is, the free space between the tray or the vial layer and the upper shelf (m); and β is a coefficient, that holds 0.001 for symmetric profiles and uniform flow distribution but can change with shelf vertical location when considered.

The previous correlation can also be used to calculate the pressure difference between two points along the vapor streamlines, moving away from the maximum values, if the distribution of the vapor is sufficiently uniform with respect to the different sides of the shelf. This can be useful, for example, to evaluate the operating conditions of vials in different positions over the shelf, in order to estimate the variance of the batch; instead of considering L_s, the distance y must be used in the formula. These conclusions agree with theorical and computational results and experimental data reported elsewhere (Zhang and Liu 2012; Ganguly et al. 2017; Sane et al. 2017).

9.3.2 MAXIMUM PRESSURE DIFFERENCE AND COORDINATES FOR MAXIMUM PRESSURE

If the flow distributes uniformly in the different directions, the maximum overpressure over the shelf and the maximum pressure difference along the shelf are very similar, and they can be easily calculated by the previous correlation, as discussed.

As an example, Figure 9.2c shows the data relative to the bottom shelf of the industrial apparatus; it can be seen that the maximum overpressure (that is, the increase with respect to the reference pressure at the point of maximum pressure over the shelf, which is relevant for evaluating the maximum product temperature in a vial) and the maximum pressure variation over the shelf (that is the difference between the minimum and maximum pressure, which is affecting the variance between vials) are very close. Differences smaller than 0.2 Pa are evaluated, while in the worst conditions in the upper shelves the difference becomes large: the maximum overpressure is still of the same order, but pressure differences along the shelf can reach 3.5–5 Pa at the maximum sublimation rate ($J_w = 1$ kg h^{-1}m^{-2}).

The maximum overpressure is strongly affected by the shelf clearance, as shown in Figure 9.2d, but remains anyway very small. It can be observed that by decreasing the clearance, the loading capacity of the apparatus increases as the maximum overpressure and generally the pressure over the shelf; but the batch uniformity also increases because the difference between different points is reduced.

Very different is the situation on the upper shelves, and in particular on "duct shelves", that is, the shelves at the level of the duct axis (see shelf 12 in Figure 9.2b); in this case the overpressure is very limited, but a much larger pressure difference can be observed, as a consequence of the strong pressure reduction in correspondence of the duct inlet. This is evident looking in Figure 9.2a with the pressure evolution along the shelf, in the chamber lateral clearance, and in the duct. Figure 9.4 (upper graphs) shows the distribution of vapor velocity and pressure in the chamber

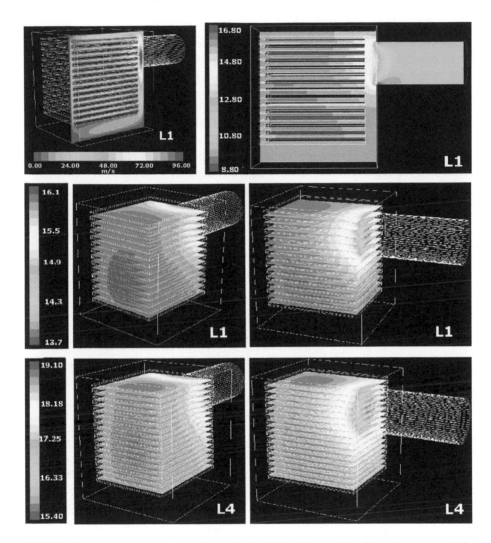

FIGURE 9.4 Upper graphs: Absolute velocity field (left graph, m/s) and pressure (right graph, Pa) in the vertical plane passing for the duct for the configuration L1. Middle and bottom graphs: contour plots of the absolute pressure (Pa) over each plate for two configurations with different free clearance: L1, 14 shelves, $s = 67$ mm; L4, 17 shelves, $s = 44$ mm. Front (left) and back (right) view; J_w 1.0 kg $h^{-1}m^{-2}$, operating pressure 10 Pa.

(a vertical plane passing for the duct axis is considered), demonstrating that the strong depression is caused by the large increase in the gas velocity in the lateral clearance zone just in front of the duct.

Middle and bottom graphs compare the pressure distribution over the different shelves by varying the shelf clearance.

CFD simulations must be carried out on the geometry considered to evaluate pressure drop in these cases. The results can be generalized, in order to obtain correlations useful for design; a possible procedure to determine the two additional parameters required is discussed in the following text.

In principle, Equation (9.11) can still be applied, provided that the fraction of flow in the considered direction can be properly estimated, as the distribution of the flow over the different plates changes and is not uniform anymore. An example of this type of plot is given in Figure 9.10 in the open access paper by Barresi and Marchisio (2018). Alternatively, the β coefficient must be evaluated for each shelf, to include the duct-effect correction: for the shelf in correspondence with the lateral duct, which shows the strongest variation, the pressure drop coefficient is 1.6–2.0 times that evaluated for the undisturbed profile, depending on the shelf clearance.

The second parameter that must be estimated is the location of the pressure maximum, $y_{max.}$, which is not the center of the shelf anymore.

Figure 9.5 shows an example for the considered industrial apparatus: the maximum pressure and its location depend on the shelf position and on the clearance and

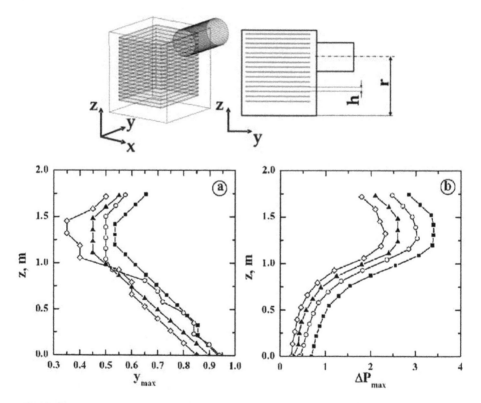

FIGURE 9.5 Variation in the location of the maximum of the pressure value along the shelf in the y direction, y_{max} (a), and maximum pressure variation, ΔP_{max} (b), as a function of the z coordinate for the four configurations of the large-scale industrial freeze-dryer, at 1.0 kg h^{-1}m^{-2} sublimation flux: $\Diamond$, L1 (h = 110 mm); $\blacktriangle$, L2 (h = 101 mm); $\bigcirc$, L3 (h = 93.5 mm); $\blacksquare$, L4 (h = 87 mm), with 43 mm constant product slab. Reprinted from Barresi et al. (2018), with permission from Elsevier. In the upper part the geometry of the apparatus is shown.

only weakly on the sublimation rate (especially for the duct shelf). As the variation is regular, empirical design correlations can be easily obtained from the CFD data. An example is given in Figure 11 in Barresi and Marchisio (2018).

Once y_{max} is known, if the side far from the duct is taken as origin of the coordinate axis, the maximum overpressure can be calculated considering the pressure increase, with respect to the reference pressure, along the y distance; the minimum pressure can be estimated considering the pressure variation, from P_{max}, along $(L_s - y_{max})$. This must be considered as an estimation, as some deviations occur from the predicted behavior close to the edge.

9.3.3 ESTIMATION OF THE INTRAVIAL VARIANCE DUE TO PRESSURE GRADIENTS

To evaluate the intravial variance (in terms of maximum experienced product temperature, drying time, residual moisture, etc.) it is possible to use a variance estimation tool, based on a two-scale approach, that allows accounting for not only the pressure gradients but also the effect of radiation from walls and windows and of temperature differences along the shelf.

The off-line multiscale model of which an example of results is shown in Figure 9.6, couples a CFD model of the freeze-drying chamber with a monodimensional model for the primary drying product (vial model). The latter describes the evolution of the

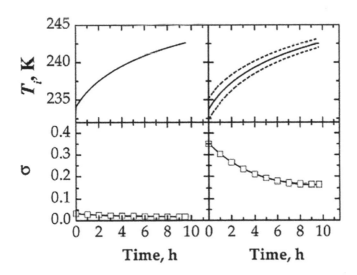

FIGURE 9.6 Time evolution of the mean value of interface temperature, T_i (upper graphs, solid line), and of the standard deviation, σ (lower graphs), for the vials on the 1st (left side) and on the 12th (right side) shelf (configuration L3). Dashed lines identify the upper and lower bound of the interface temperatures in the vials. Solution of bovine serum albumin; frozen product thickness = 7.2 mm, shelf temperature = 258 K, reference chamber pressure = 10 Pa. Reprinted from Barresi et al. (2018), with permission from Elsevier.

ice interface, of the product temperature and can estimate the remaining moisture content, using as boundary conditions the pressure values obtained from the CFD simulation. Thus, many vials corresponding to the slightly different operating conditions experienced by the product in different parts of the equipment can be modeled, to recover the information on the batch variance. It can be noted that the variance decreases with process time, but very different values are obtained for the shelves positioned differently with respect to the duct.

As discussed before, it is possible to use the CFD data to obtain semiempirical correlations for the pressure distribution in the chamber, to be used in the dual scale model for design purposes, as shown by Rasetto et al. (2010).

The different possible dual-scale approaches have been discussed by Barresi et al. (2018), who have also shown the relevance of temperature differences in the fluid along the shelf; a good design must consider also the maximum load in the apparatus, as conditions can be significantly different from those occurring during qualification tests, carried out at no load. It was shown that in some cases pressure gradients might be counterbalanced by a proper design of the heat fluid path.

On-the-fly multiscale models, with a simple zero-dimensional vial model implemented in the CFD code, may be used to obtain more detailed CFD simulations of the freeze-dryer chamber, to consider the effect of nonuniform sublimation rate in the different part of the apparatus.

9.4 ESTIMATION OF THE CHAMBER RESISTANCE AND DUCT ENTRANCE EFFECT

Figures 9.2 and 9.4 demonstrate that pressure drops along the drying chamber are generally small, while a significant pressure drop can occur in the entrance zone to the duct, owing to large velocity variations and the developing of a new radial velocity profile in the duct.

Obviously, in case of duct in the bottom, pressure gradients in the vertical direction may increase, causing eventually a significantly different value of the average pressure over the shelves: in this case the batches would be more uniform if each shelf is considered, but differences between shelves can become larger. The size of the lateral clearance plays also a relevant role; it must be enough to avoid large pressure drops, and in case the position of the duct is not symmetric, the best design would consider different lateral clearances, to ensure equal flux in each one. This would guarantee minimum pressure drops and uniform flow distribution over the shelves. A good design criterion would be to have similar pressure drops over the shelves and in the lateral clearance, to minimize total pressure variation and maximize uniformity. In fact, uniform flow distribution can be obtained also by increasing the resistance of the shelf zone, but at the cost of increasing the pressure over the product. Thus, the charts for the flow distribution over the shelf discussed in the previous section can also be used to evaluate the design of the clearance between the shelf pack and the wall.

To allow an evaluation, the two main contributions must be considered separately. The first considers the variation from the "reference pressure", which would correspond to the value measured by a well-located pressure gauge (that is, one located far from inert gas inlet or condenser duct) and the chamber exit. Once calculated, the actual pressure at the inlet of the duct, the pressure drop in the duct inlet can be estimated. Obviously, the best solution would be to model the whole apparatus, because it has been shown that pressure drop is very sensitive to geometric details and even sharpness of the entrance profile. Anyway, some data on the effect of geometry and on the entrance length for the velocity development profile are available (Barresi and Marchisio 2018).

The results obtained from the CFD simulation of the whole chamber are strongly specific for the geometry considered; thus, they are strictly valid only for the case investigated. But it is possible also in this case to obtain generalized correlations, which will have a similar structure, and use it to build design charts for a class of equipment. An example, for the same industrial apparatus, is shown below. As the volume occupied by the shelves is the same in all four configurations considered, a generalized correlation can be recovered.

The variation from the front zone (taken as reference) and the point along the median axis of the shelf, in correspondence with the duct entrance, has been considered.

The chamber pressure drop is inversely proportional to the chamber pressure; the average pressure should be considered for a more accurate estimation, but for calculation purposes it is easier to use the reference pressure, as shown in Equation (9.12):

$$\Delta P P_{ref} = a_1 \dot{m}_w^{\,2} + a_2 \dot{m}_w, \tag{9.12}$$

where $\dot{m}_w$ is the total mass flow rate (kg/h), corresponding to the product of sublimation flux times the shelf surface times the number of shelves. The coefficients in Equation (9.12) depend on the design of the chamber and duct, but Figure 9.7 (upper graph) shows that they remain the same for the different shelf configurations; and it can be expected that the same type of semiempirical correlation may be employable for other geometries, provided the right coefficients are calculated from relative CFD data.

The bottom graph in Figure 9.7 is an example of the design charts that can be easily plotted using the previous correlation for the different shelf configurations. The pressure drop has been calculated for different reference pressures and a wide range of sublimation rates (of course, the plot refers to a specific shelf size). The filled symbols are the data calculated by CFD, which correspond to a fix value of pressure at the duct outlet (10 Pa), and thus to a different chamber pressure for any sublimation rate.

9.5 EVALUATION OF BUTTERFLY AND MUSHROOM VALVE CONDUCTANCE

The conductance of the duct can be significantly modified by the presence of a valve, even if qualitatively the behavior is similar to that of an empty duct. To complete design or verification of the performance of the freeze-dryer, the real conductance of the duct and valve set must be evaluated, considering the actual pressure at the exit of the chamber estimated in the previous step.

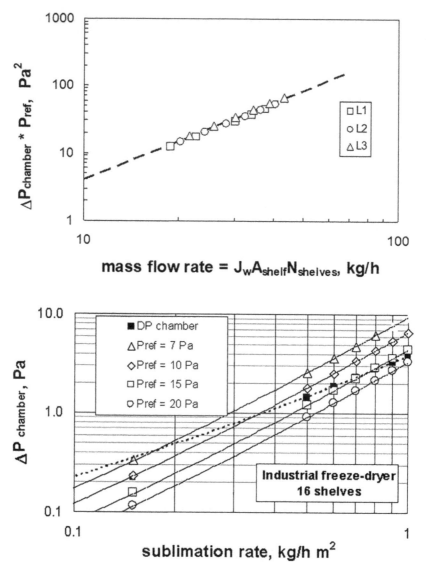

FIGURE 9.7 Upper graph: Chamber pressure drop in the industrial freeze-dryer shown in Figure 9.5: generalized correlation using the reference pressure, validated by CFD data obtained with three different shelf configurations. Lower graph: Design chart for the chamber pressure drop, for different reference pressures; L3 configuration. The filled symbols are the original data calculated by CFD, which correspond to a fixed value of pressure at the duct outlet (10 Pa) and thus to a different chamber pressure for any sublimation rate.

In case a butterfly valve is used to isolate the condenser, the significant reduction of the conductance due to the presence of the disk may be explained in part with the reduction of the section available and with the obstacle caused by the disk itself. But as the pressure drop in the duct is mainly related to the inertia effect due to the developing of the velocity profile, in the butterfly valve the increase may be explained with the larger modification of the velocity profiles.

Figure 9.8a–b shows the geometry considered with two different disk shapes and an example of the velocity field obtained by CFD simulation.

The whole set of results obtained in choked flow conditions for the two different disk geometries is shown in Figure 9.9. The simulations have been carried out for DN 700, but the results have been generalized referring to the mass flux. All the plots and correlations refer to the use of surface average pressures. A linear relationship exists between the inlet (chamber) pressure and the critical mass flux also in this case, like in an empty duct. Thus Equation 9.2 can be used also to describe the pressure–critical mass flux relationship, with a proper value of the coefficients.

It is evident that the conductance of the two valves is significantly different: the geometry corresponding to the flat disk allows larger vapor flow rates. The slope of the curves is not only different for the two geometries (and therefore the parameters

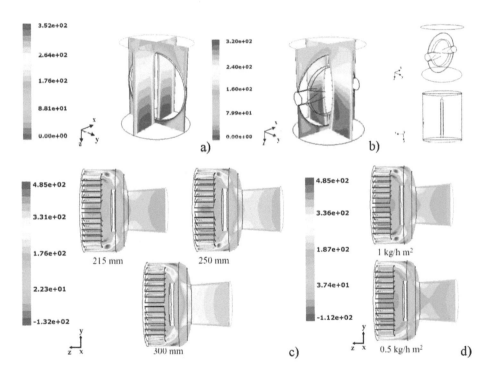

FIGURE 9.8 Upper graph: Butterfly valve. CFD results for the flow-through butterfly valve with different disk shape: (a) simple flat disk and (b) complex profiled disk. Velocity magnitude (m/s). Boundary conditions: low pressure slip, inlet pressure 36.5 Pa. The geometry considered in the CFD simulations, to obtain the data shown in Figure 9.9, is shown in the upper right corner. Bottom graphs: CFD results for the mushroom valve at the entrance of the condenser. (c) Axial velocity on plane $x = 0$ for different values of the valve distance; sublimation rate = 1.0 kg h^{-1}m^{-2}. (d) Axial velocity on plane $x = 0$ for different values of the sublimating rate; $l_{valve} = 215$ mm. A few plots previously published in Barresi and Marchisio (2018).

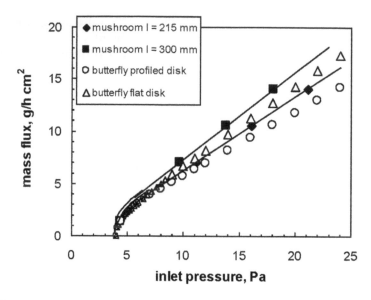

FIGURE 9.9   Comparison of mushroom and butterfly valve conductance; $P_{out} = 4$ Pa. From Barresi and Marchisio (2018).

in Equation 9.2 will be different), but it is also different from that of the empty ducts, especially in the case of the profiled valve (compare Figure 9.1c).

$$\text{Flat disk} : P_{in} = 1.31\ j_w + 1.28 \tag{9.13}$$

$$\text{Profiled disk} : P_{in} = 1.65\ j_w + 0.55 \tag{9.14}$$

The slope is higher in the case of the valve; thus, the concept of equivalent length can be used only for very rough estimates, as the equivalent length would increase with the inlet pressure considered.

From 6 to 40 Pa, the equivalent length of the flat disk valve would increase from about 4 to about 12, while for the profiled disk valve it would increase from about 6 to about 40.

By using the design equations previously proposed, describing respectively the choked flow, the subcritical, and the transonic region (also for these conditions a similar relationship holds for the valve and for the ducts), it is possible to realize design charts also for valves. An example is shown in Figure 9.10.

The design charts for the butterfly valves are similar to those already presented for empty ducts, showing the variation of the mass flux with the inlet and outlet pressure. As in that case, two different plots can be realized: in the first one the influence of the outlet pressure at a given inlet pressure is evidenced; in the second one the outlet pressure is used as a parameter, and the effect of the inlet pressure is evidenced.

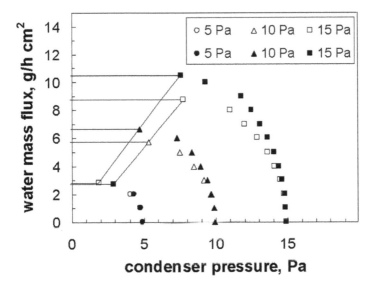

FIGURE 9.10 Design chart for butterfly valves: mass flux vs. outlet pressure, for different inlet pressures. Different disk shapes: open symbols, profiled disk; filled symbols, flat disk. Calculated from DN 700 simulations, with no-slip and adiabatic wall boundary conditions. Reprinted from Marchisio et al. (2018), with permission from Elsevier.

The resistance due to the valve is always very significant; as the duct is generally short in normal equipment, in practice the flow conductance is determined by the valve. Thus, the design chart of the valve can be reasonably used to predict the behavior of the whole line.

Figure 9.8c–d shows the velocity field and the geometry for a horizontal condenser connected with a DN 700 conical duct and equipped with a flat disk mushroom valve (the restricted section has a 700 mm diameter, while the valve disk has a 750 mm diameter); the results have been generalized using the mass flux also in this case. Equation 9.2 can be used also in this case, but the constant α_p depends on the valve disk position:

$$P_{in} = 1.56 \left(\frac{4 l_{valve}}{d} \right)^{-0.5} j_w + 1.2, \tag{9.15}$$

where l_{valve} is the valve distance from the vessel bottom (mm) and d is the valve disk diameter (mm).

Figure 9.9 shows the conductance of the mushroom valve (for different disk positions), comparing it with that of the butterfly valves.

ACKNOWLEDGMENTS

Valuable contributions from Miriam Petitti, Valeria Rasetto, Davide Fissore, Daniele Marchisio (Politecnico di Torino) are gratefully acknowledged.

Figure 9.1c, some plots of Figure 9.8 and Figure 9.10 from Barresi and Marchisio (2018) https://doi.org/10.1016/j.ejpb.2018.05.008, under Creative Commons Attribution License (CC BY) https://creativecommons.org/licenses/by/4.0/

REFERENCES

Alexeenko, A. A., Ganguly, A., and S. L. Nail. 2009. Computational analysis of fluid dynamics in pharmaceutical freeze-drying. *J. Pharm. Sci.* 98:3483–3494.

Barresi, A. A., and D. L. Marchisio. 2018. Computational Fluid Dynamics data for improving freeze-dryers design. *Data Brief* 19:1181–1213.

Barresi, A. A., Fissore, D., and D. L. Marchisio. 2010a. Process Analytical Technology in industrial freeze-drying. In *Freeze-drying/lyophilization of pharmaceuticals and biological products*, ed. L. Rey, and J. C. May, 3rd rev edition., Chap. 20, 463–496. New York: Informa Healthcare.

Barresi, A. A., Pisano, R., Rasetto, V., Fissore, D., and D. L. Marchisio. 2010b. Model-based monitoring and control of industrial freeze-drying processes: Effect of batch nonuniformity. *Drying Technol.* 28:577–590.

Barresi, A. A., Rasetto, V., and D. L. Marchisio. 2018. Use of Computational Fluid Dynamics for improving freeze-dryers design and understanding. Part 1: Modelling the lyophilisation chamber. *Eur. J. Pharm. Biopharm.* 129:30–44.

Batchelor, G. K. 1965. *An introduction to fluid dynamics.* Cambridge: Cambridge University Press.

Fissore, D., Pisano, R., and A. A. Barresi. 2015. Using mathematical modeling and prior knowledge for QbD in freeze-drying processes. In *Quality by design for biopharmaceutical drug product development*, AAPS Advances in the Pharmaceuticals Sciences Series 18, ed. F. Jameel, S. Hershenson, M. A. Khan, and S. Martin-Moe, Chap. 23, 565–593. New York: Springer Science+Business Media.

Ganguly, A, and A. A. Alexeenko. 2012. Modeling and measurements of water – vapor flow and icing at low pressures with application to pharmaceutical freeze-drying, *Int. J. Heat Mass Transfer* 55:5503–5513.

Ganguly, A., Alexeenko, A. A., Schultz, S. G., and S. G. Kim. 2013. Freeze-drying simulation framework coupling product attributes and equipment capability: Toward accelerating process by equipment modifications. *Eur. J. Pharm. Biopharm.* 85:223–235.

Ganguly, A., Varma, N., Sane, P., Bogner, P., Pikal, M. J., and A. Alexeenko. 2017. Spatial variation of pressure in lyophilization product chamber part 1: Computational modelling. *AAPS PharmSciTech* 18:577–585.

Kshiragar, V., Tchessalov, S., Kanka, F., Hiebert, D., and A. Alexeenko. 2019. Determining maximum sublimation rate of a production lyophilizer: Computational modelling and comparison with ice slab tests. *J. Pharm. Sci.* 108:382–390.

Marchisio, D. L., Galan, M., and A. A. Barresi. 2018. Use of Computational Fluid Dynamics for improving freeze-dryers design and process understanding. Part 2: Condenser duct and valve modelling. *Eur. J. Pharm. Biopharm.* 129:45–57.

Maxwell, J. C. 1879. On stresses in rarified gases arising from inequalities of temperature. *Phil. Trans. Royal Soc. London* 170:231–256.

Oetjen, G. W. 1999. *Freeze-drying.* Weinheim: Wiley-VHC.

Patel, S. M., Chaudhuri, S., and M. J. Pikal. 2010. Choked flow and importance of Mach I in freeze-drying process design. *Chem. Eng. Sci.* 65:5716–5727.

Petitti, M, Barresi, A. A., and D. L. Marchisio. 2013. CFD modelling of condensers for freeze-drying processes. *Sādhanā (Bangalore) – Acad. Proc. Eng. Sci.* 38:1219–1239.

Pisano, R., Fissore, D., and A. A. Barresi. 2011. Heat transfer in freeze-drying apparatus. In *Developments in heat transfer*, ed. M. A. dos Santos Bernardes, Chap. 6, 91–114. Rijeka, Croatia: InTech. Open access book, available at www.intechopen.com/books/show/title/developments-in-heat-transfer.

Rasetto, V. 2009. Use of mathematical models in the freeze-drying field: Process understanding and optimal equipment design. PhD diss., Politecnico di Torino.

Rasetto, V, Marchisio, D. L., Fissore, D., and A. A. Barresi. 2010. On the use of a dual-scale model to improve understanding of a pharmaceutical freeze-drying process. *J. Pharm. Sci.* 99:4337–4350.

Sane, P., Varma, N., Ganguly, A., Pikal, M. J., Alexeenko, A., and R. H. Bogner. 2017. Spatial variation of pressure in the lyophilization product chamber part 2: Experimental measurements and implications for scale-up and batch uniformity. *AAPS PharmSciTech* 18:369–380.

Searles, J. 2004. Observation and implications of sonic water vapour flow during freeze-drying. *Am. Pharm. Rev.* 7:58–69.

White, F. 2009. *Fluid mechanics*, 7th Edition. New York: McGraw-Hill.

Zhang, S., and J. Liu. 2012. Distribution of vapor pressure in the vacuum freeze-drying equipment. *Math. Problems Eng.* 2012: Article ID 921254, 10 pp.

Zhu, T., Moussa, E. M., Witting, M., et al. 2018. Predictive models of lyophilization process for development, scale-up/tech transfer and manufacturing. *Eur. J. Pharm. Biopharm.* 128:363–378.

Index

Note: *Italicized* page numbers indicate a figure on the corresponding page. Page numbers in **bold** indicate a table on the corresponding page.

Milton Keynes UK
Ingram Content Group UK Ltd.
UKHW022041141024
449569UK00015B/687